BEES &
BEE-KEEPING

BEES &
BEE-KEEPING

DEREK HALL

CHARTWELL
BOOKS, INC.

Published in 2010 by
CHARTWELL BOOKS, INC.
A division of BOOK SALES, INC.
276 Fifth Avenue
Suite 206
New York
NY 10001
USA

**Copyright © 2010 Regency
House Publishing Limited**
The Red House
84 High Street
Buntingford, Hertfordshire
SG9 9AJ, UK

For all editorial enquiries, please contact
Regency House Publishing at
www.regencyhousepublishing.com

ISBN-13: 978-0-7858-2629-3

ISBN-10: 0-7858-2629-7

Printed in China

**This book is intended primarily to stimulate
interest in the subject of bee-keeping and is no
substitute for obtaining expert advice on the
subject before proceeding further.**

CONTENTS

INTRODUCTION

Bees are remarkable little insects that have been around for about 100 million years or so, going quietly and efficiently about their business of gathering their food from the nectar and pollen produced by flowers. In doing so, bees perform a valuable and vital service to the plants they visit and,

BEES & BEE-KEEPING

OPPOSITE LEFT: A bee sculpture, a symbol of the Fraternal Hive, Alumni House, Georgia Tech.

OPPOSITE RIGHT & ABOVE: As well as benefiting themselves, bees perform a vital service to the plants they visit, pollinating them and ensuring the continuation of plant species.

ultimately, to human beings and other animals, for they aid pollination through their foraging activities and thus ensure the spread and continuation of plant species.

Some bees, such as bumblebees and honeybees, are termed social bees, for they have evolved an incredible and intricate system of task-sharing and a well-defined hierarchy whereby hundreds or even thousands of the creatures are able to live together in close harmony, building the internal structures of their home, protecting

LEFT: A giant bumblebee sculpture, composed of plant material on a metal frame, located in the gardens of the Eden Project, Cornwall, England, the home of the world's largest bionome.

ABOVE: Bee street art in a telephone call box in California.

and tending to one another's needs, and laying in food stocks to tide the colony over in leaner times. This very food – honey – is another reason why bees are so important to us human beings, for the precious substance has been proved to have nutritional and medicinal benefits over thousands of years. Today,

ABOVE & RIGHT: Most wild bee species construct either single or complex nests in the ground, but others make earthen, leaf or resin (propolis) nests on rocks and plants, at times utilizing crevices in trees, rocks or plant stems, and even using insect borings or plant galls.

unlike our ancestors, we no longer raid bees' nests to steal their honey, but

The products of the beehive are various, present not only in foodstuffs, health and beauty products, but also extend to such practicalities as candle- and soap-making.

'farm' bees in a highly regulated manner, temporarily opening, rather than destroying, the specially constructed nests we provide for them, and taking only some of the honey, leaving sufficient behind to satisfy the bees' own needs.

The general principles of bee-keeping have not changed significantly in a couple of hundred years, although a multitude of hive designs have gradually become available, together with new tools of the trade, foodstuffs, and other associated paraphernalia that inevitably

accumulates around such activities. Opinions differ where the minutiae of bee-keeping are concerned, though the basic rules continue to hold true whatever the method or methods

Honey is much more than a delicious treat: since ancient times it has been valued for its important nutritional and healing properties.

employed: understand the way honeybees live and work with them; use good-quality tools and good practices; take sound advice but, as you become more experienced, modify it to suit your own needs and those of your bees; be mindful of how neighbours and family can be affected by the presence of bees; keep tools and

equipment clean and in good condition; only work with good bees obtained from a reliable source.

Today, eager to play a part, however small, in helping to maintain our world's diminishing bee stocks, while at the same time enjoying all the delicious rewards, many of us are making the decision to keep bees

ourselves. Here, many aspects of bees and bee-keeping are discussed, but this book is intended to be no more than an introduction to the world of bees and a simple guide to caring for them. The more serious among you would benefit from other, more specialized books on the subject, where the finer points of the craft are discussed in much more detail. It is also worth approaching one of the various bee-keeping organizations to discover, at first hand, the more precise practicalities of what is involved.

LEFT: A bee-keeper, clad in protective clothing, is inspecting his bees. Beehive frames are the structural elements that hold the honeycomb or brood comb within the hive body or 'super'. They are a key part of the modern 'movable' hive, since they can readily be removed to inspect the bees or to extract excess honey.

OPPOSITE: Skeps, or basket hives made from coils of grass or straw, were once common in Europe. These have a single entrance at the base, with no internal structure provided for the bees. Skeps have two disadvantages: bee-keepers cannot inspect the interiors for pests and diseases, and honey removal often means destroying the entire hive.

BEES & BEE-KEEPING

OPPOSITE: The swarm of giant golden bees, an art installation decorating the northern entrance of the Eureka Tower, Melbourne, Australia, and which seem ready to buzz away at any time. Eureka Tower, the world's tallest residential tower, which opened in October 2006, is already establishing itself as an architectural icon.

LEFT: Following reports of dwindling numbers of bees worldwide, this sculpture on Governors Island, New York, was intended to focus public attention on the importance of the continuing presence of honeybees to the future of the planet.

CHAPTER ONE
BEES & OTHER INSECTS

Think of the word 'bee', and most of us will conjure up the image of a small, plump, rather fuzzy creature, buzzing languidly from flower to flower on a hot summer's afternoon, pollinating the flowers in our gardens as it proceeds. This picture is a perfectly accurate one, and the creature we are seeing in our mind's eye is most likely to be the bumblebee or humble-bee (Bombus genus), or even the honeybee (*Apis mellifera*). But the bumblebee and

LEFT: There are over 250 species of bumblebee in the northern hemisphere.

ABOVE: A honeybee on blossom.

OPPOSITE: The cuckoo bee (Nomada genus) has an unusual way of resting by hanging from a plant by means of its mandibles (jaws).

honeybee are only two of the many different types, or species, of bee to be found throughout the world, species which altogether number at least 20,000 in all. Some bees, including bumblebees and honeybees, live together in social colonies, while others, such as mining bees and cuckoo bees, live a more or less solitary existence. While it is more usual to imagine bees living harmoniously together in hives or colonies, there are, in fact, many more types of solitary bees in existence on the planet.

All bees, whatever their species, are part of the vast assemblage of tiny creatures known as insects. Insects include some of the most familiar

BEES & OTHER INSECTS

animals on Earth, as well as some of the least-known and most obscure. Among the best-known are the various types of flies, wasps, fleas, beetles, moths, butterflies, dragonflies, grasshoppers and aphids. The

caterpillar-like creatures found by gardeners on plants or in the soil may be the larval stages of many different types of insects, while other insects are either so small or so rare or secretive that only those dedicated to their study ever

Bees belong to the successful insect group containing a vast number of creatures in various guises, and which include the tiger swallowtail butterfly, *Papilio glaucas* (above), and the spur-throated grasshopper, subfamily Melanoplinae (opposite).

encounter them, and even then possibly only by chance.

Insects are remarkable in many ways, and they are arguably the most successful creatures on earth. There are several million different insect species – we aren't sure exactly how many, because new ones continue to be discovered in remote places, such as rainforests – and because there are so many different types they have been able successfully to colonize and exploit almost every habitat on land apart from the very coldest parts of polar regions. Yet surprisingly, despite their supreme

OPPOSITE: The yellowjacket hoverfly (*Milesia virginiensis*) mimics the hornet, the adult feeding on pollen.

RIGHT: A wasp is typically defined as any insect of the order Hymenoptera and suborder Apocrita that is neither a bee nor an ant.

ability to insert themselves into every available niche on land, only a very tiny number of species have ever found a way to live in the sea. To get an idea of the sheer scale of insects in the natural scheme of things, consider these facts: there are many, many times more insect species than every other species of animal in the world put together. And if it were possible to place all the insects living in the world on one side of a colossal pair of scales, placing every other creature on the other side, the insects would easily outweigh all the others.

Insects are invertebrates – animals without backbones – and they belong to a vast assemblage of creatures that live variously in the air, in the soil, on trees, under rocks and in all kinds of habitats in fresh water and in the sea. Worms, snails, slugs, starfish, jellyfish, spiders and mites are all invertebrates, although there are thousands of other kinds.

BEES & OTHER INSECTS

Insects belong to the biggest subdivision of the invertebrates, known as the arthropods (Arthropoda), which is the grouping that also includes crabs and spiders among others. The bodies of arthropods are encased in hard outer coverings known as exoskeletons, and their limbs are jointed (in fact, the word arthropod comes from the Greek for 'joint-limbed'). Arthropods may be thought of as wearing a suit of armour with movable joints; the exoskeleton protects the animal's soft internal body parts and helps it retain water inside the body. Insects themselves are placed within the class Insecta, the biggest of several classes within the phylum Arthropoda, with bees grouped into a further subdivision known as the order Hymenoptera. This order, one of 28 or so in the class Insecta, also contains the wasps and the ants – the bees' closest relatives (and in the case of wasps their probable ancestors). Finally, bees are grouped into various families.

LEFT & OPPOSITE: Snails (Mollusca) and starfish (Echiderms), together with the Arthropoda, are phyla (principle taxonomic categories) within the invertebrate group, which includes 95 per cent of all animal species.

BEES & BEE-KEEPING

The classification of the honeybee is as follows:

Kingdom: Animalia (all living animals)
Phylum: Arthropoda (insects, crabs, spiders, mites and others)
Class: Insecta (insects)
Order: Hymenoptera (bees, wasps, ants and sawflies)
Suborder: Apocrita (bees, wasps and ants)
Section: Aculeata (non-parasitic bees, wasps and ants)
Superfamily: Apoidea (bees)
Family: Apidae
Genus: Apis
Species: mellifera

Even at the species level there may be further subdivisions. For example, the honeybee exists in a variety of additional forms known as races or hybrids. Races may be distinguished from one another by slight morphological variations, but often the designation into another race is decided purely on geographical distribution. Thus the East African race of honeybee is a slightly smaller – and considerably more aggressive (perhaps

OPPOSITE: Andrenidae belong to one of the four non-parasitic bee families that contains some species that are crepuscular or active only at dusk or in the early evening.

BELOW: A European or Western honeybee collecting nectar from auricula flowers.

more defensive is a more accurate term) – version of the European honeybee (*Apis mellifera*), and is classified as *Apis mellifera scutellata*.

Apart from an exoskeleton, insects also have wings – the only group of invertebrates to possess them – and they have therefore mastered the sky. This is another reason for their overwhelming success in reaching and exploiting new habitats as well as providing them with a means of escaping enemies. Most insects have two pairs of wings, but some, such as beetles, have only a single pair, and a few species (for example, parasites like fleas and lice as well as the primitive silverfish) have dispensed with wings altogether during the course of their evolution. Most arthropods have sensory feelers, or antennae; insects have only one pair of antennae, although there are other arthropods that have more. Finally, all insects have three pairs of legs.

CHAPTER TWO
BEES IN PARTICULAR

Anatomically, bees are typical insects in as much as they have many of the general features of the class as a whole. Much of the information in the general description that follows relates to the honeybee, although many other bee species have similar characteristics. The body, clad in its protective exoskeleton made of a tough substance known as chitin, is divided into three major parts: the head, the thorax and the abdomen. The body is composed of compartments called segments. These are clearly visible in the larval forms of many insects such as fly maggots or butterfly caterpillars, where they appear as a series of rings running around the body. In adult insects, however, the segments are usually less clearly defined and often – as in the head region – become fused together and impossible for the casual observer to discern. The best way to see the segmentation is to look at the flexible abdomen of an adult insect such as a bee, where the ringlike arrangement is often still apparent. The body of a bee is usually covered in tiny hairlike structures that give many species a fuzzy or bristly appearance, and which help to gather pollen and also aid regulation of the body temperature.

THE HEAD AND SENSES

The head bears various sensory organs. First, there are the antennae, which are usually composed of 13 segments in male bees, 12 in females. Insects such as bees use their antennae to probe and feel objects at close quarters, such as when constructing the combs within the hive or communicating with other bees in the darkness of the nest or hive. Bees generally live inside places such as tree cavities, under compost heaps, or in man-made hives where there is little or no light, making the senses of smell and touch, rather than sight, extremely important in these environments. Thus most of the communication that takes place between bees in the hive is through touching antennae. The head also houses a brain, which consists of a collection of about a million nerve cells known as neurons, a nervous system

OPPOSITE: The head of a honeybee, showing the antennae, compound eyes and ocelli, and the mandibles (jaws).

LEFT: The head and thorax of a bumblebee.

PAGE 32: A bumblebee (Bombus genus).

PAGE 33: A honeybee in close-up.

that allows the brain to communicate with, and send messages to, the rest of the body.

Bee antennae carry thousands of sensory organs, some of which are designed to detect touch, some to detect odours and others to detect tastes. They also 'hear' airborne sounds at close range, thanks to an ability to sense the motion of air particles through their hairlike touch receptors. A bee about to enter a hive or colony will be 'smelled' by guards near to the entrance to make sure that the incoming member has the correct odour of that particular colony. If not, the guards will force it away. A virgin queen bee also produces a sex pheromone that is detected by male bees when she is on her mating flight. Experiments using honeybees have shown that they can also detect flavours such as sweet, sour, bitter and salt, just as we can.

A bee has two types of eyes and like those of other insects these differ

LEFT & OPPOSITE: As well as its mouthparts, used for chewing, grasping, cleaning and handling pollen and wax, a bee also has a tubelike device, known as a proboscis (a complex 'tongue') that enables it to obtain nectar from flowers.

greatly from human eyes. The first type are known as compound eyes, and a bee has two of these – one on each side of the head. Compound eyes are large and obvious relative to the size of a bee's head, and typically appear as dark, shiny, oval structures. Compound eyes are composed of many tiny six-sided facets, each of which is essentially an individual lens. Each of these lenses has only a narrow field of vision, but the images from adjacent ones overlap, and the closer the bee is to an object the sharper the image that is produced. Experiments suggest that a bee's vision is sharp for a distance of up to about 3 feet (1 metre).

Bees also have three smaller ocelli, or simple eyes, on the tops of their heads, two at the side and one at the front in a triangular configuration. (*Ocellus* is a Latin word meaning 'little eye'.) 'Simple' is a rather misleading word in this instance, however, because the ocelli are anything but simple in their construction. Although the ocelli have a light-gathering function and a wide field of view, just like the compound eyes, it is thought that they do not form images and that their job is to assist the bee in maintaining stability

while in flight. The ocelli are well-adapted to measure the changes in the perceived brightness of the insect's surroundings as it moves its body during flight. Other theories pertaining to the role of ocelli include possible use as light adaptors or for sensing polarized light.

As well as seeing most of the wavelengths of light in the visible range that we humans can also see, bees are capable of seeing ultraviolet light, which is invisible to us, although they cannot detect the colour red. The bee is capable of navigating by ultraviolet light, which even penetrates cloud cover. Some flowers have what are known as nectar guides; these are patterns that guide pollinators to the nectar store within the flower. Some flowers have nectar guides that are only visible in ultraviolet light, ensuring that only bees or other ultraviolet light-detecting insects will pollinate them.

The mouthparts of a bee are carried at the front of the head, the size and shape of which vary from species to species. The jaws, or mandibles, at the sides of the mouth, are used for grasping and chewing objects, cleaning other bees, and manipulating pollen

BEES IN PARTICULAR

and wax. As well as these chewing mouthparts, bees also have a tubelike device, known as a proboscis, that ends in a glossa or tongue. The tip of the bee's tongue is sometimes spoon-shaped, with spines on the top and sides, while in other cases it is quite long, depending on the species. The main function of the proboscis is for sucking up nectar as well as liquids such as water and honey. It is also used to transfer food from one bee to another, and can be folded out of the way when not in use. Muscles in the bee's head pump fluids up the proboscis, from whence they enter the pharynx and then the oesophagus. Bees ingest (take in) as well as egest (expel) food, so the pump can also work in reverse when they need to pass food to the developing larvae or feed other workers.

Other mouthparts include the labium, a single structure formed from two fused secondary maxillae or feelerlike palps. It can be described as the floor of the mouth, and, with the maxillae, helps to manipulate food during mastication.

RIGHT & OPPOSITE: Though difficult to see, the thorax is in three parts and is the area to which the wings and legs are attached.

The head also carries salivary glands to moisten food. Queen bees have a special gland above the jaws, known as the mandibular, which they use to secrete a pheromone (which can be likened to a chemical message) known as queen substance; the presence of this pheromone, at the appropriate concentration, maintains the social organization within the colony. This gland is very reduced in size in drones (male bees). Brood food

glands, also called hypopharyngeal glands, produce the important food substance known as royal jelly.

THE THORAX

The next part of a bee's body – the thorax – is made up of three segments, but as in the head these are not always easy to see in the adult insect. The thorax is the part of the body to which the wings and legs are attached. The two pairs of wings are membranous structures supported by veins in a clearly visible network, the membranous nature of the wings being the reason for the scientific name of the

order Hymenoptera. Large flight muscles, running from the top to the bottom of the exoskeleton within the thorax, contract and relax, pulling the wings up and down to power them in flight at up to 230 beats a second; the wings are folded back along the top of the body when the bee is at rest. In bees, the hind pair of wings is shorter than the front pair, and is often scarcely visible, especially when the animal is at rest. The leading edge of each hindwing bears a row of tiny hooks called hamuli, that fit into grooves on the trailing edges of each forewing. This arrangement effectively couples the hind and forewings together so that they form a large flight surface, beating in unison when in flight. Once the bee is back in the confined area of the hive, it can 'unzip' the pairs of wings and fold them up so that they occupy less space and avoid damage.

The three segments of the thorax each bear one set of jointed legs ending in pads and claws. As well as enabling the bee to walk about, each pair of legs is designed for a particular task related to the bee's lifestyle. In honeybees the

The wings are folded back along the top of the body when the bee is at rest.

front pair of legs is used to clean and groom the head, eyes and mouth, while a special notched cleaning apparatus helps them groom the antennae. The central pair of legs is also used to help clean the body, and the legs are also employed for tasks such as loosening pollen from the pollen baskets and cleaning the wings. The hind pair of legs is specially designed for the collection of pollen, in that each leg is flattened, and on worker bees is covered with long, fringed hairs resembling a brush. Pollen grains collected from flowers stick to the hairs of the bee's body and are brushed backward by the legs, pressed into a pellet that is then packed into a hollow section of the leg known as the pollen basket. The collected pollen usually can be seen as a thick yellowish band around the hindlegs of a bee that is out foraging among flowers.

THE ABDOMEN

The abdomen is the 'end section' of an insect and the part where the segmentation of the body is most clearly to be seen in the adult form. The abdomen carries few appendages, compared with the head and thorax, but it is the place where many of the bee's internal body organs are located. Insects breathe through small pores called spiracles, located on the surface of the abdomen and also the thorax. These connect with an internal system made up of a series of tubes, known as tracheae, as well as air sacs that carry oxygen to the internal organs. Within the abdomen is a tubelike digestive system that includes a crop or honey sac.

The underside of a worker bee's abdomen has wax glands, used for making the wax cells in the hive or nest that are required for storing the bees' food and raising their young. The disclike scales of wax exude onto four pairs of wax pockets, also on the underside of the abdomen. On the dorsal (top) side of the worker bee's abdomen, near to the tip, is the Nasonov gland, which secretes a sweet-smelling pheromone that helps to orientate and guide other bees back to

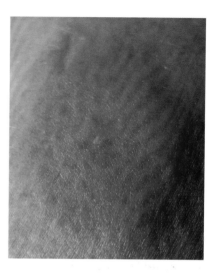

the hive. The gland is exposed when the abdomen is flexed, and can be seen in action when bees land at the hive entrance after a foraging trip. The emission of the pheromone is accompanied by rapid wing-beating that helps to disperse the odour to incoming bees. This is only one of several ways in which bees communicate with one another through the use of pheromones.

THE STING

The most significant appendage on a bee's abdomen is the sting, the bee's stinging apparatus being its method of defence. In some species, the sting is only rarely used, while in others it is either absent altogether or is so small as to be rendered useless against aggressors. Some species have a reputation for stinging aggressively and en masse, and such an attack can lead to the death of the victim. Finally, there are some bee species that do not sting at all.

The sting is actually a modified ovipositor, or egg-depositing apparatus, and as such is only found in the female caste members, i.e., the queen and workers. The sting, located at the tip of the abdomen, is a narrow, needlelike structure, often with barbs on its outer surface, and it is connected to a venom gland and venom sac inside the abdomen. The barbs on the queen's sting are less pronounced than those on the workers', and she uses her sting

BELOW & OPPOSITE: The pollen basket, or corbicula, is part of the tibia on the hindlegs of the four related lineages of apid bees that used to comprise the family Apidae, and which include honeybees, bumblebees, stingless bees and orchid bees.

only against rival queens. It is widely known that wasps are able to sting repeatedly, but some bee species – for example, some types of honeybee – sting only once and then die. The reason for this is because the worker bee's sting is very heavily barbed, and when the bee attacks a marauder, such as a mammal intent on raiding the nest, the barbs penetrate deeply into the victim's skin. When the bee moves away, the sting gets dragged from her body and is left behind in the victim, continuing to drive venom into the wound by the independent action of its muscles. The action of stinging releases from the bee a scent called an alarm pheromone, that alerts other bees to come to the defence of the hive or nest so that they, too, may sting the attacker.

DEALING WITH BEE STINGS

The bee's venom contains a variety of chemicals that can destroy the victim's cell tissues. These include enzymes and

peptides that break down the fat layer lining the cell. The venom also attacks the victim's immune system.

Following a bee sting, the victim's body releases a chemical called histamine. This encourages the blood vessels to dilate, allowing the body's defences to get to the site of the attack and neutralize the venom. On occasions, however, an allergic reaction to the venom can set in, when the overdilation of the blood vessels leads to a drop in the body's blood pressure, causing the cells to stop receiving oxygen and therefore making breathing difficult. This is known as anaphylactic shock, which can lead to spasms and death if not treated quickly.

Happily for most people, however, a single bee sting is a painful but not life-threatening experience, and a hazard to be expected by those whose lives centre around bees, be they amateur gardeners, professional horticulturalists or bee-keepers in particular. Of course it depends on where you are stung; stings on the ears, back of the neck, nose and other sensitive regions will be much more unpleasant than, say, the back of the hand, where there are fewer sensitive nerve endings. But as we have seen, the

Two honeybees using their probosces to take nectar from a flower.

sting, even minus the bee that delivered it, will go on doing its work for several minutes once embedded in the skin, pulsating under its own muscle system as it pushes deeper into the skin and forces more venom into the unfortunate victim. The best course of action is to remove the entire sting as soon as possible, using a pair of fine tweezers. An ice pack will help to reduce the subsequent swelling, while applying a proprietary cream or lotion intended for the relief of bee stings may also help, but carefully check the instructions first. Multiple stings may require a visit to a doctor or hospital outpatients' department to check that no allergic reaction is likely to set in (see above).

BEE DIGESTION

Food – mainly nectar and pollen from plants in the case of most bees – is taken into the mouth, then passes along the alimentary canal (a thin tube running through the thorax), and into the honey sac or crop in the abdomen. When the bee takes nectar from a flower, the nectar passes into the honey sac to be taken back to the hive.

BEES & BEE-KEEPING

Digestive juices break down the nectar and pollen in the honey sac, which has a valve at one end that allows some of this food to pass into the large intestine to be used by the bee's own body. Once in the large intestine, further enzymes and digestive juices release proteins.

Nutrients then pass through the walls of the large intestine into the bee's body cavity to be utilized. The remains of the food then pass into the small intestine. Here a series of thin threadlike organs, known as the malpighian tubules, remove waste products.

OPPOSITE: Swarming is a natural process whereby the queen bee leaves her colony, accompanied by a large group of worker bees, to form a new colony, leaving a new queen to take over in her place.

BELOW: Honeybees drinking from a bird bath.

45

CHAPTER THREE
HOW BEES EVOLVED

As already mentioned, there are thousands of different bee species throughout the world today. Bees probably evolved about 100 million years ago in the middle of the Cretaceous Period, at around the same time that flowering plants first appeared on earth. The Cretaceous Period was the time in Earth's history

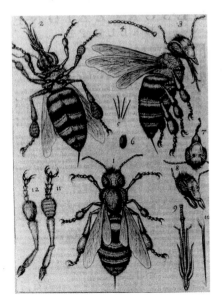

when dinosaurs still reigned supreme, but the newly emerging birds and mammals were beginning to appear in increasing numbers. Prior to the advent of the flowering plants, most of the existing large land plants reproduced by relying on the wind to blow pollen produced in cones from one plant to

the cones of another. The male pollen from one cone then fertilized the female egg from another cone, producing a seed. This is exactly the method still used by many plants today, and includes conifers such as pines and larches.

Then, in the Cretaceous Period, some plants began to bear specialized structures as a means of reproduction. These were flowers, and unlike conifers, which rely on the wind to blow their pollen about, flower-bearing plants used other agents to help disperse their pollen from one flower to the next. They mostly relied on animals for this task, especially insects, and to ensure that the insects visited their flowers they provided sweet, sugary nectar as a source of highly nutritious food. The flowers also evolved bright colours and other devices to help attract the insects to them. But to obtain the nectar, the insect needed to be able to delve inside the flower to reach the nectar glands at

BEES & BEE-KEEPING

pollination to be effective it requires not only a favourable wind but also the production of vast amounts of pollen, since much is wasted when it is blown about. This is inefficient for the plant. The insect-pollinating method is much less wasteful, since pollen is inadvertently carried directly from flower to flower. In time, flowering

its base. While doing so, pollen from the flower's anthers (part of the male flower structure) would brush against the insect's body and stick to it, and when the insect then visited another flower of the same species, some of the pollen on its body would rub off onto the stigma (part of the female flower structure) inside the flower. From here the pollen would make its way to the female egg, eventually fertilizing it and producing a seed.

This method of reproduction clearly had advantages. For wind-blown

plants even began to evolve particular flower shapes and other strategies that ensured only certain types of insects visited them, thus still further reducing the chances of wasting pollen.

At about the same time, bees began to differentiate themselves from their insect ancestors, which many scientists believe may have been a species of wasp, such as those of the family Crabronidae. Instead of feeding on other small animals, as is the way of wasps, the newly evolved bees became herbivores, feeding on nectar and pollen instead.

Today, the role of bees in human food production cannot be underestimated. It is calculated that about a third of our world food supply relies on insect pollination, the majority of which is carried out by bees, particularly the Western honeybee. In California, pollination of the almond crop, which takes place early in the year, depends on the activities of more than one-third of all

OPPOSITE & RIGHT: Flowers, such as this lemon blossom and purple clover, gradually evolved more elaborate ways of attracting insects and making sure the insects left bearing the maximum coverage of pollen.

HOW BEES EVOLVED

LEFT & OPPOSITE: About a third of our world food supply relies on insect pollination, the majority of which is carried out by bees.

BELOW: The Ophrys genus has some species, such as the bee orchid, that look and smell so like female bees that males flying nearby are irresistibly drawn in to attempt to mate with the flower, thus dispersing the pollen.

the domesticated honeybees in the United States. Consequently, large numbers of hives are therefore transported into the area so that pollination can be carried out successfully. Similar movements of bees take place when apple blossom is pollinated in states such as Michigan and Washington.

HOW FLOWERS HELP BEES

It is as much in the interest of a flowering plant to have a bee or other suitable insect visit it as it is for the bee to visit the flower, since the plant relies on such visits to propagate its particular species. The most obvious external features of typical insect-attracting flowers are their colour and their scent, since both help the bee to detect the flower from a distance. Some of the orchids (such as the bee orchid)

Native plants, particularly the older varieties of perennials, are usually best for native bees, and can be used in both wild areas and gardens.

have evolved petal shapes or other flower parts that closely resemble the females of the insects they wish to attract. When a male insect lands on the flower in a vain attempt to mate with it, it naturally helps to pollinate the flower in the process. Some flowers have special 'landing stages', making it easier for insects to approach the flower, and a few have devices that only trigger access into the flower when the correct sort of insect lands. The long, tubelike petals of the flowers of white dead-nettle are a method of ensuring that only the 'correct' type of bee pollinates it; in this instance the bee in question is the red carder bee (*Bombus ruderarius*), a type of bumblebee with a long tongue designed to penetrate deep down inside the flower to reach the nectar.

LEFT: The white dead-nettle (*Lamium album*) is designed to attract the red carder bee, which has a tongue that is long enough to penetrate its depths.

OPPOSITE ABOVE LEFT & RIGHT: The sourwood tree and sourwood honey from North Carolina.

OPPOSITE BELOW: The manuka flower produces a honey with legendary qualities.

OPPOSITE: Bees play a vital part in pollinating plants, such as apple trees in an orchard, for no fruit will form unless the flowers are first fertilized by cross-pollination.

THIS PAGE: Honey comes in different colours, depending on the plant source, i.e., white to very light amber in the case of cotton and orange, and dark with a distinctive menthol flavour in the case of eucalyptus.

As we have seen, bees feed on pollen and nectar. The pollen covers the male flower parts, the anthers and, to a lesser degree, also the filamentous stamens that support the anthers. The bee gathers up this light, dusty substance as it moves about inside the flower, in the process inevitably transferring some pollen it has already

BEES & BEE-KEEPING

Flowers have bright markings and attractive fragrances to attract bees and other insects to them so that pollination will take place. Some also have dark lines known as 'honeyguides', which are believed to help insects find their way into the flowers themselves.

Much as wine is affected by the terroir, or the complete environment in which the grapes are produced, such as soil, topography and climate, so honey with different tastes, smells, colours and textures will also result from the types of flowers visited by bees, especially when one particular plant predominates in an area, i.e., heather, clover, orange blossom, etc. Other honeys, however, can be delicious blends of many different floral sources such as wildflowers.

HOW BEES EVOLVED

collected from other flowers to the female parts. The sweet, sugary nectar is produced in special flower glands called nectaries, located at the base of the perianth. Plants that are less discriminating when it comes to the insects that visit them usually have easily accessible nectaries, whereas those that are more specific in their choice may have nectaries that are accessible only to the insects adapted for reaching them.

BELOW: A wild nest suspended from a tree.

OPPOSITE: A honeybee on a foraging expedition.

BEES & BEE-KEEPING

To some, a field of rapeseed is a glorious sight, the acid yellow flowers producing blocks of colour on a spring landscape, while to others it may herald the onset of the dreaded hay fever season.

Rapeseed (*Brassica napus*), however, besides being a valuable oil-producing crop, is a heavy nectar producer, and honeybees produce a honey from it that is light in colour with a slight peppery taste. The honey must be extracted immediately after processing is finished, otherwise the honey will quickly granulate in the honeycomb making it impossible to extract. The honey is usually blended with milder honeys, if for table use, or sold as a bakery grade sweetener. Rapeseed growers often contract with beekeepers to ensure the pollination of the crop.

CHAPTER FOUR
THE MAIN TYPES OF BEE

Thousands of different bee species are found around the world today, and the hive-dwelling, flower-pollinating behaviour adopted by the honeybee is only one of many different lifestyles exhibited by them; there are even 'grades' of sociability among the social bee species, ranging from loose gatherings of similar bees to systems in which the nest or hive runs according to a distinct hierarchical order, and which consists of a matriarchal queen, workers (sterile females) and drones (males), each with well-defined roles. Other bees live solitary existences, and some indulge in practices such as parasitism. Behaviour such as nest-building also takes many forms, and the names of mining bees and leafcutter bees provide strong clues as to the methods used by these species to

RIGHT: A swarm of Western honeybees.

OPPOSITE: A buff-tailed bumblebee.

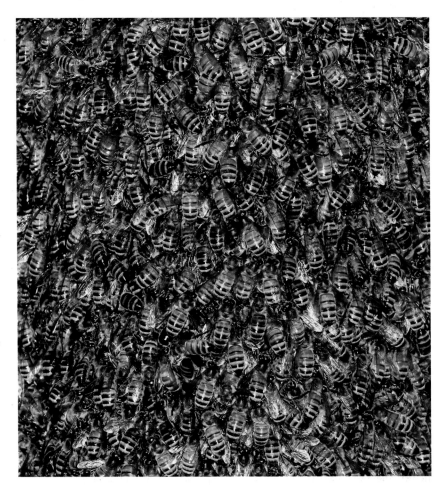

construct their homes. Not all bees fly in bright daylight; some families of bees contain species that fly at dawn or dusk when light levels are much lower, in which case their ocelli are enlarged and that much more sensitive. Many of these species are adapted to pollinate flowers that only open at night or in low light, such as evening primroses. Some bees live in regions such as deserts, where flying in the searing heat of the day is best avoided, and they also pollinate flowers at night.

Although many bees conform to the slightly rounded shape typical of species such as bumblebees and honeybees, there are others that look distinctly wasplike in shape or colour, for example, the leafcutter bee (*Megachile willughbiella*) and the species known as *Nomada fulvicornis*. These species have narrow waists and black-and-yellow abdomens, which enhance their 'waspishness' that much

THE MAIN TYPES OF BEE

The furry, colourful bumblebee, its fat body a nutritional store, is a welcome addition to any backyard or country garden.

Bumblebees (Bombus genus) are social insects that can be distinguished by their black and yellow body hairs, often arranged in bands. Some species, however, may have orange or red on their bodies, or may be entirely black. Another obvious (but not unique) characteristic is the soft nature of the hair, this consisting of long, branched setae, known as pile, that covers the entire body, making the bee appear fuzzy in outline. Bumblebees are best distinguished from similarly large, fuzzy bees by the form of the female hindleg, which is modified to form a corbicula, this being a shiny concave surface that is bare but surrounded by a fringe of hairs used to transport pollen (in similar bees, the hindleg is completely hairy, and pollen grains are wedged into the hairs for transportation). Sadly, some bumblebee species are now under threat of extinction.

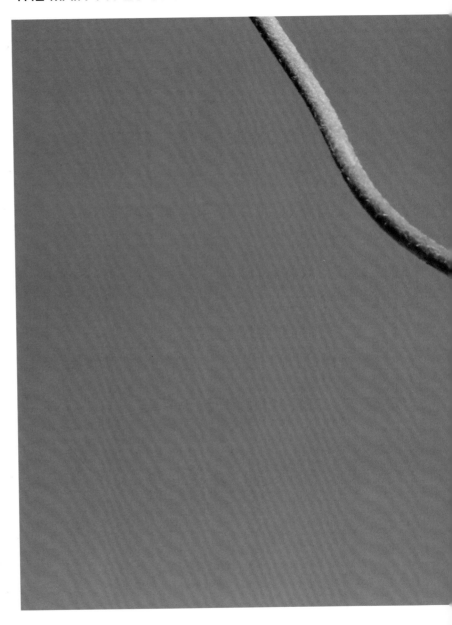

THE MAIN TYPES OF BEE

The cuckoo bee (above) and leafcutter bee (opposite left) are narrower at the waist than honeybees and bumblebees, giving them a more wasplike appearance.

more. The abdomen of the cuckoo bee (*Nomada fulvicornis*) terminates in a sharply pointed end. This species is almost all black in colour, making it look even less like a typical bee. The

carpenter bee (*Xylocopa violacea*) is another unusual-looking bee, this being a handsome, harmless insect with shiny blue-black body parts and attractive violet wings.

BEES & BEE-KEEPING

FAR RIGHT: This tiny little halictid (sweat bee) is most common in North America.

FAR RIGHT BELOW: The beautiful blue-black carpenter bee has violet wings and is quite harmless.

The world's smallest bee is a dwarf species called *Trigona minima*; it measures about 0.08-in (2.1-mm) long and is one of the species of stingless bees. At the other end of the scale, the largest bee in the world is an Indonesian leafcutter bee called *Megachile pluto*,

which attains 1.5in (38mm) in length. The most common species of bee worldwide is the European honeybee, our old friend *Apis mellifera*, but in North America the most common kind is a type of halictid or 'sweat bee'

(Augochloropsis species), whose common name comes from the fact that they often land on people's skin to lick up the salt from their perspiration. This can be an alarming experience, since they often resemble wasps in appearance. Sweat bees also have an unusual way of obtaining pollen, known as buzz pollination, in which the bee grasps the anther in its jaws and vibrates its wings, causing the pollen to be dislodged onto its body.

SOLITARY AND COMMUNAL BEES

Solitary bees are most abundant and diverse in places such as the deserts of the Mediterranean basin and North America's south-west. After mating, the female builds a nest from glandular secretions, subsequently laying down provisions of pollen and nectar for her offspring to eat when they hatch. While doing this, solitary bees play an important part in pollination, and

LEFT & OPPOSITE: The tawny mining bee (*Andrena fulva*) is a solitary bee which nests in burrows in the ground.

THE MAIN TYPES OF BEE

The mason bee (*Anthidium florentinum),* left, and the leafcutter bee, opposite, are the most commonly seen members of the family Megachilidae.

many species specialize in pollinating only certain types of flower. This means that both the bee and the flower are closely dependent on one another for their species' survival. The nest of a solitary bee varies according to the species; for example, it may be a few cells built inside a tree hollow or a hole in a bank. In the case of the tawny mining bee (*Andrena fulva*), the nest is often a hole dug in a lawn with a mound of soil near to the entrance resembling a small volcano.

Another interesting type of nest-building behaviour is seen among members of the family Megachilidae, the most common members of which are the leafcutter and mason bees. Instead of using glandular secretions to build their nests, they collect materials such as mud, chewed leaves, flower petals and pieces of animal fur. Gardeners are often bemused to discover semicircular chunks bitten out of the edges of rose leaves, the work more often than not of leafcutter bees such as *Megachile centuncularis*. Once

BEES & BEE-KEEPING

THE MAIN TYPES OF BEE

LEFT: This mason bee (*Osmia ribifloris*) has a beautful blue iridescent sheen. It is one of several species referred to as blueberry bees, and is a Megachilid native to the coastal mountains of southern California.

RIGHT: A leafcutter bee visiting a man-made nest.

the bee has bitten a suitable piece of foliage from the leaf, she carries it away between her legs before chewing it into a paste ready for applying to the nest.

Although solitary females each make their own individual nests, some species make nests close to one another, and this concentration of similar bees can sometimes give the impression that they are in fact social insects. Large groups of solitary bee nests are known as aggregations. Another form of association occurs when females of particular species share a common nest site – a large cavity in a tree, for example – but each makes separate provisions for her own cells. An advantage of this type of arrangement is that the nest site only needs to have a single common entrance, and it is therefore easier to defend from parasites and marauding predators.

SOCIAL BEES

These are bees that live together in a community, the most advanced type of which are known as eusocial, and are found among the bumblebees, the stingless bees, and their relatives the honeybees. In these systems, each colony has a queen and a large number of female bees called workers. At certain times of the year, the colony also produces drones (male bees) for the purpose of mating and producing new queens.

BUMBLEBEES

Bumblebees (Bombus species) are found all over the world, but tend to favour higher altitudes and latitudes than many other bee species, although some lowland tropical species also exist. One of the reasons that bumblebees are found in cooler climes is because they are thought to be able to regulate their own internal body temperatures. Big, fuzzy and familiar for the droning or humming sound they make as they go from flower to flower,

RIGHT: A bumblebee on buckwheat.

OPPOSITE: A bumblebee sharing a flower head with a honeybee.

THE MAIN TYPES OF BEE

BEES & BEE-KEEPING

Bumblebees form annual colonies, and it is only the mated queens that survive the winter to begin new colonies the following spring. Then, the queen builds a nest and prepares and provisions it ready for raising her young. The bumblebee nest is often constructed under the ground, for example, in a disused mouse's nest, or perhaps in the dense base of tall grasses or under a compost heap in the garden. It is a ball of grass and moss with wax cells inside. At the peak of the season, in mid to late summer, there may be

bumblebees are without barbed stings, so they can sting more than once. There are over 250 species of bumblebees worldwide, and they are characterized by their yellow- or orange-and-black body hairs, often arranged in bands. They feed on nectar and gather pollen to feed their young. Bumblebees are important natural pollinators, but their numbers have declined in recent years.

OPPOSITE: A bumblebee on love-in-the-mist.

ABOVE: A female *Bombus terrestris*.

RIGHT: The charming teddy bear bee (*Amegilla bombiformis*) is native to Australia.

between 50 and 200 individuals in the nest consisting of the queen and workers. The nests are rarely perennial, and new ones are built each year. At the end of the season, unfertilized queens and drones are produced. These mate, and the old queen, the workers and the drones then die, while the young queens disperse to find overwintering sites until they, too, will start new colonies in the spring.

ABOVE: A bumblebee nest.

RIGHT: A bumblebee taking nectar.

OPPOSITE: An Australian blue-banded cuckoo bee (genus Thyreus).

CUCKOO BEES

The cuckoo bees (Psithyrus species) are now regarded by some authorities as a type of bumblebee, and have been given subgenus classification status. Cuckoo bees, like the bird with the common name, are parasitic, in this case laying their eggs in bumblebee colonies. To do this, a female cuckoo bee first enters an existing bumblebee nest and kills the resident queen. The cuckoo bee is able to accomplish this 'assassination' because she is more powerful and more venomous. When the cuckoo bee larva hatches, it first eats the food originally intended for the host's larva, and then eats the larva itself. The worker bumblebees then feed and raise the young of the cuckoo bee instead of their own young. Female cuckoo bees do not have the pollen-collecting structures seen on other bees, since they themselves have no intention of providing food for their young.

STINGLESS BEES

The 20 or more genera of stingless bees are related to the honeybees, bumblebees and carpenter bees. The stingless bees, which we are concerned with here, belong to the tribe Meliponini, but to add to the confusion there are other types of bees that do not sting; the Meliponini, furthermore, are not actually devoid of a stinging apparatus – it is simply that it is too small to be an effective defence weapon. Stingless bees are found in subtropical and tropical regions such as South-East Asia, Australia, Africa, Mexico and South America. Many

species are prized for their honey, and are kept and tended in a similar way to Apis honeybees.

Stingless bees are social insects, nesting in hollow tree trunks, beneath the ground in holes, in rock crevices, and in a variety of man-made objects such as old garbage cans and storage drums. Although they do not sting, they nevertheless defend their nests vigorously, either biting with their sharp jaws or, in the case of a few species, by emitting a mandibular secretion that causes painful blistering. Hives of some species can be extremely large, with anything from a few hundred to about 80,000 individuals living inside. The bees keep their pollen and honey in egg-shaped pots made from beeswax, which are arranged around brood combs in which the larvae are kept and reared.

OPPOSITE ABOVE: The cloak and dagger cuckoo bee (genus Thyreus) preys on the nests of blue-banded bees.

OPPOSITE BELOW: A cuckoo bee of the family Andrenidae.

RIGHT: The stingless bee *Meliponula ferruginea*.

THE MAIN TYPES OF BEE

THE MAIN TYPES OF BEE

About 14 of the wild bee species in Australia are stingless. All are small, black insects with hairy hindlegs adapted for carrying pollen. They are popular in gardens since they help pollination, do not pose a threat to humans, and can be used for small-scale honey production. Since the bees produce little excess honey, however, care must be taken to ensure their own stocks are not depleted to the point where the colony is at risk of starvation. The stingless bees of Central America were kept for their honey as far back as the Mayan period, and were considered sacred. Today, however, they are endangered through a combination of habitat loss caused by deforestation and the use of insecticides to control other insect pests, as well as by changes in bee-keeping practices that came about with the introduction of the Africanized *Apis mellifera* honeybee.

HONEYBEES
There are about seven species of honeybee in the world and just over 40

RIGHT: Honeybees arrived in North America with the first colonists from Europe.

OPPOSITE: The Western honeybee.

84

THE MAIN TYPES OF BEE

LEFT: Giant honeybees (*Apis dorsata*) are to be found mainly in south and South-East Asia.

BELOW: A dwarf honeybee. Though small, the nest is usually quite conspicuous, being round or elongated in shape.

OPPOSITE: The Italian honeybee (*Apis mellifera ligustica*) is commonly found in southern Europe and North and South America.

subspecies, all grouped within the genus Apis. Today, however, whenever the term honeybee is used, we usually mean the domesticated European or Western honeybee, *Apis mellifera*. Honeybees as a group probably originated in southern Asia, but various races became well-established in areas such as Africa and Europe, both in the wild as well as in domestic hives. Fossils of honeybees first appear in deposits from the Eocene-Oligocene period, about 35 million years ago. They have been long prized by humans for their honey-producing properties, and were kept at least as far back as the time of the ancient Egyptian pharaohs several thousand years ago.

Dwarf honeybees, such as *Apis florea* and *Apis andreniformis*, are little bees which build small, usually quite conspicuous nests, often round or elongated in shape, among tree and shrub foliage, the nests measuring about 8-in (20-cm) across. The bees have only a tiny stinging apparatus on their abdomens. Dwarf bees are found in southern and South-East Asia in places such as Thailand, while giant honeybees, such as *Apis dorsata*, are common in the same sorts of regions of the world as dwarf bees. They build conspicuous nests that may be 3ft (1m) or more in diameter, usually high up in the branches of trees, on cliff sides or even on buildings. Unlike dwarf honeybees, which can often be handled

with minimal risk of being seriously stung, giant honeybees are much more formidable defenders of their nests when confronting raiders (including human beings).

Although, when domesticated, *Apis mellifera* are kept in hives, they build nests in cavities when in the wild. There are a number of different *Apis mellifera* subspecies around the world, especially in Europe, the Middle East and Africa.

Subspecies originating in Europe include the following:

Italian bee (*Apis mellifera ligustica*). This the most common variety of honeybee found especially in southern Europe and in North and South America. The worker has a yellow body with brown or blackish stripes. Drones are a gold colour without stripes. The

queen can easily be identified thanks to her large, golden-orange abdomen. The Italian bee also has a long tongue, meaning it can probe into deeper flowers than shorter-tongued varieties.

Coming originally from the warm climate of the Italian peninsula, the bee favours fine weather and can be a prolific breeder. It is able to find flowers growing for much of the year,

so there is ample opportunity to store stocks of nectar and pollen for the winter. In cooler climates, however, when natural food becomes scarce during winter, the bee still needs plenty

BEES & BEE-KEEPING

OPPOSITE: Carniolan bees from Slovenia, a subspecies of the European or Western honeybees.

BELOW: A Caucasian honeybee (*Apis mellifera caucasia*) taking up water from a wet stone.

of food – more so than other types of bees in temperate regions – since it maintains large hive populations. The Italian bee is calm and gentle on the comb and keeps its hive clean, making it easy to maintain and manage. As long as it has plenty of space it is not readily given to swarming. This is a prodigious producer of honey given the right conditions.

Carniolan bee (*Apis mellifera carnica*). This bee originated in the Carniola alpine region of Slovenia. Due to its

THE MAIN TYPES OF BEE

OPPOSITE: The dark European honeybee *(Apis mellifera mellifera)* is also known as the German honeybee.

BELOW: A European or Western honeybee on a nectar-gathering expedition.

Balkan origins, it is hardy and accustomed to quite long winters. Also in winter the hive population falls significantly before it builds up again in the following season (this is due to a shortage of natural food in winter in its original homeland). At this time, therefore, it can be maintained on lower food levels than, for example, the Italian bee. But it can also react quickly once the weather warms up, rapidly increasing the hive population and swarming. The Carniolan worker bee is a calm and gentle type, and quite dark in colour – usually grey or black with grey abdominal stripes. Its long tongue allows it to feed on flowers such as red clover. It also swarms readily. Drones have black abdomens. Queens are all black in colour. Not only is the Carniolan tolerant of the needs of hive management, but it also produces little propolis (see page 127).

Caucasian bee (*Apis mellifera caucasia*). Originally from the Caucasus Mountains of Eurasia, this variety is usually regarded as very gentle and industrious, and quiet on the comb when being examined. It has been reported, however, that when these bees are crossed with other strains, some of the resulting colonies have proven problematic. Caucasian bees react to winter conditions by reducing their hive populations and feeding sparingly on their stored food reserves. The bees have very long tongues. Workers are dark grey with lighter stripes on the abdomen. Drones and queens are dark in colour.

Dark European bee (*Apis mellifera mellifera*). Also called the German honeybee, this old variant is the 'type specimen' used in the classification by the eminent taxonomist Carolus Linnaeus in 1758. It is a small, dark-coloured bee.

There are also other European variants, such as *Apis mellifera iberiensis* from the Iberian peninsula, *Apis mellifera cecropia* from Greece and *Apis mellifera ruttneri* from Malta.

Subspecies originating in Africa include:

African honeybee (*Apis mellifera scutellata*). The African honeybee is smaller than any of the European bees and produces smaller cells. It occurs around West, Central and East Africa but has also hybridized in South America, Central America and the southern United States. The African honeybee is a hard-working variety that is active early in the day and still gathering from flowers late on into the evening. It is considered to be an aggressive species that reacts quickly and defensively in the event of the colony being disturbed, tending to attack intruders en masse. Major disturbances may cause the colony to swarm.

Cape bee (*Apis mellifera capensis*). This dark-coloured bee originated in the coastal regions of South Africa. Workers can produce fertile eggs that develop into other female bees, although they are not able to mate with drones. In the absence of a queen, the workers can produce the 'queen substance'.

There are other African variants, such as *Apis mellifera adansonii* from Nigeria, *Apis mellifera lamarckii* from Egypt and Sudan, and *Apis mellifera jemenitica* from Uganda, Somalia, Sudan and Yemen.

In addition, there are various subspecies of bees that come from other parts of the world, such as Asia and the Middle East. These include *Apis mellifera meda* from Iraq, *Apis mellifera anatolica* from Turkey and Iraq, *Apis mellifera syriaca* from Israel and *Apis mellifera pomonella* from central Asia.

OPPOSITE: The African honeybee is smaller than its European counterpart. Here the queen is marked with a red spot to distinguish her from the others.

ABOVE: A honeybee with its pollen basket clearly visible.

CHAPTER FIVE
THE LIFE OF THE HONEYBEE

Before bee-keeping is even contemplated, it's important to know how the honeybee lives its life, who's who in the hive, how the hive hierarchy works, what are the natural enemies of bees, and so on. This will make it much easier for you to manage and care for your bees, be aware of any likely problems, and finally harvest the all-important honey that will be produced. The honeybee lives a remarkable life; to the casual observer, a quick glimpse into a hive may appear to show little more than a mass of bees, endlessly moving about but achieving little else. Yet these fascinating creatures not only display a level of cooperation, communication, self-sacrifice and tireless toil not seen in any other invertebrate, but also attain a higher level of social organization than many more advanced animals.

We can begin by looking at the different types of bee in the colony. In the wild, honeybees make their nest inside some form of natural cavity, such as a hollow tree or in an abandoned animal nest hole in a bank. When this occurs, the nestful of bees is termed a colony. In the case of domesticated bees, the colony simply lives within a man-made structure (the hive).

THE QUEEN

The queen is the matriarch of the colony. Although she does not actually boss the other bees in the hive about, as her regal-sounding title would suggest, she nevertheless maintains behavioural control over the colony through the

secretion of hormones. Under almost all conditions, apart from those pertaining to Cape bees, for example, the queen lays the eggs that will develop into other individuals in the colony. That is her prime role, and to enable her to accomplish this efficiently, she is cosseted, fed and constantly attended to by the other members of the hive (the workers). Each colony has only one queen at a time, and she is recognizably bigger than the other bees (about 0.6–0.75-in /15–19-mm long) with a long, tapering abdomen; the large abdomen is a necessity, since for almost her entire life the queen is little more than an extremely efficient egg-laying machine. She may produce 1,000 to 2,000 or more eggs in a single day, and since she may live for two or three years, she is capable of laying over 2 million eggs in her lifetime. Like other

OPPOSITE: Carniolan honeybees with the larger queen in the centre.

ABOVE RIGHT: An older queen larva in a queen cell lying on top of a wax comb.

FAR RIGHT: A capped queen cell opened to show a queen pupa (with darkening eye).

bees in the colony, the colour of the queen varies according to the species.

BORN TO BE QUEEN
It will be necessary to produce a new queen at some point in the life of a colony. The reason for this may be because the current queen is failing to lay sufficient eggs, or she may be injured or may even have died. A new queen is also produced when the colony becomes too big for its nest or hive, triggering a swarm. This is a mass flight from the nest of the old queen and about half of the workers to form a new colony. The new queen stays with

the remainder of the colony in the original hive.

The new queen-to-be starts her life as a fertilized egg laid by the incumbent hive queen; at this stage the egg is identical to all the other eggs destined

from egg to adult, and 24 days in the case of a drone completing the same cycle.

Initially, the colony produces as many queens as it needs to guarantee replacements; sometimes 20 or so may develop, all at slightly different stages. When this happens, the first queen to emerge sets about killing all the other potential royal rivals. She does this by biting into the cells containing the other queens and stinging them to death. Occasionally several queens will emerge from their cells at the same time, when a battle ensues until only one queen is left victorious. The new queen then feeds herself, or is fed by workers, for a few days as she continues to mature. Then she will undertake a series of orientation flights in the vicinity of the hive to prepare herself for the mating flight. During these exercises, the virgin queen familiarizes herself with landmarks so that she can navigate back to the hive following the mating flight that is to come.

to be workers. The fertilized egg is placed in a queen cell; this is larger than the cells used to rear workers and is orientated vertically on the comb as opposed to worker and drone cells that are arranged horizontally. Here the egg develops for about three days, after which time the egg's outer covering dissolves and releases a small, grublike larva. Now, worker bees immediately begin to feed the larva by bringing food to its cell. There are countless visits by the workers bearing a rich, highly nutritious food called royal jelly that is composed of pollen, honey and enzymes. For the first two or three days, all new larval bees receive this royal jelly. Indeed, the larvae destined to be queens get nothing else. But for the larvae which will be workers or drones, the diet becomes downgraded in both quality and quantity after a few days, with the result that they never develop into queens. At the same time, those destined to be queens begin to develop the reproductive systems and pheromone and hormone glands they will need in the future. From egg to adult queen takes about 16 days; this compares with about 21 days for a worker to develop

ABOVE LEFT: Eggs and larvae (the brood cell walls have been partially cut away).

OPPOSITE: Worker bees are non-reproductive females.

THE LIFE OF THE HONEYBEE

Once she is ready, the virgin queen leaves the hive to mate with drones (males). The queen seeks out drones from other colonies, if possible, since this prevents inbreeding and ensures that the gene pool remains vigorous. The mating flight, or sometimes several mating flights, lasting over a few days, takes place when the weather is suitable, the best conditions usually to be found being sunny afternoons. Mating flights take place around 30 to 100ft (10–30m) up in the air above fields or woodland clearings, during which time the queen emits a powerful sex pheromone to attract her suitors. She may mate with several drones, which die soon after the event, leaving their mating organs inside the queen. During her mating flights, the newly mated queen acquires enough sperm from the males she encounters to enable her to fertilize all the eggs she will ever lay in her lifetime. She stores the sperm in a sac in her abdomen known as the spermatheca. Every time she lays an egg, she will release, at the same time, a small amount of sperm from the spermatheca to fertilize it.

If, however, a prolonged spell of inclement weather prevents mating flights from taking place as planned, the queen may soon pass her mating age, and the colony will begin to prepare more queens for the task instead. If for any reason a colony is unable to produce a newly mated queen, then it will be unable to continue, in the event of which bee-keepers must make provisions for supplying another queen from other stock to ensure the survival of the hive.

Back in the hive, the mated queen begins to exert control over the hive through the emission of complex pheromones, which are mostly produced by the hypopharyngeal glands in the queen's head, near to her jaws. These emissions, known as 'queen substance' are detected by the workers and serve to unite the colony by essentially signalling to them that a queen is in residence and that all, therefore, is well. The odours and scents are passed throughout the hive. A drop in the perceived level of queen substance is a signal to the hive workers that a new queen is needed (for reasons mentioned earlier, such as a queen coming to the end of her egg-laying life) or because of overcrowding. When overcrowding occurs, the colony will swarm (see page 129).

THE WORKER

The worker is the most abundant type of bee in the hive. In fact, apart from the queen, and the drones produced when mating is required, the colony consists only of workers. Workers are the smallest hive members, ranging in size from 0.5 to 0.7in (13–17mm) or so, and they are non-reproductive females except in rare circumstances. On average, a hive may contain several thousand workers during a season. Workers do just as their name implies: they guard the nest; they do all the tidying up; they maintain the hive at the correct temperature; they make the wax cells in which the eggs will be laid; they feed and tend the queen and all the young worker larvae; they fly about on endless trips gathering food from flowers; they convert nectar into honey and store all the food they have, including the surplus honey that is collected by bee-keepers. Each worker bee is responsible for various tasks within the hive and they work in groups, although these tasks vary at

RIGHT & OPPOSITE: Worker bees collecting nectar and pollen from clover flowers. Note the pollen basket visible on the bee opposite.

OPPOSITE & ABOVE: Worker bees going to and fro in pursuit of their many duties.

PAGES 104–105: It is usually the older worker bees that go out foraging.

different stages of their lives. The first tasks of young worker bees, for example, include cleaning and polishing cells, making the hive comb and feeding larvae, whereas it is usually older workers that go on foraging expeditions. In the summer, in temperate latitudes, a worker bee lives for about six weeks; if she is born during the winter she may live for about six months.

THE LIFE OF THE HONEYBEE

A worker bee starts life as an egg laid by the queen, at which stage there is no difference between an egg destined to be a worker and an egg destined for royalty. As soon as the egg hatches into a larva, it is fed for the first three days on an identical diet of royal jelly to the one received by a future queen. Thereafter, however, the food that the worker larva gets is cut back, and as a result she fails to develop reproductive organs and various glands. After a larval stage lasting for six days, the larva turns into a pupa. Twelve days later, the fully-formed adult worker bee emerges from the pupa. To begin with, she begs food from other, more mature workers, remaining close to the centre of the hive where most of the pollen food is stored and where it is safest and warmest. Soon she is able to find her own pollen, and as she feeds she develops some of the organs needed to perform her duties as a worker. As

LEFT: A honeybee takes full advantage of a wild fennel plant.

OPPOSITE: A honeybee worker laden with pollen to bring back to the hive.

already mentioned, her first duties involve the maintenance and cleaning of the cells once the pupae have hatched and generally keeping the hive clean by removing faeces, dead bodies and other debris. Soon she moves on to other duties such as cell-building.

When the mandibular and hypopharyngeal glands in her head have developed, the worker can begin feeding larvae with a pollen and honey mixture, and later still she can feed them with royal jelly produced from her head glands. She will also attend to the queen's various needs and as she does

This man-made structure, sited in an orchard to ensure that the blossom is fertilized to produce fruit, replicates a wild beehive as far as it is possible.

so will help pass the all-important queen substance around the hive, telling other hive members that the queen is present and that all is well. But this will not occur if the queen is failing or is actually missing.

Temperature regulation is an important task carried out by workers. The optimum temperature is 95° F (35° C), and if it rises above this the workers will fan their wings to move air around and help bring the temperature down. In more extreme conditions, the workers will spread water droplets around before fanning their wings so that even cooler air is circulated. In cold conditions, the worker bees flex their body muscles to help generate additional heat to warm up the hive.

LOOKING FOR FOOD

A worker bee that is ready to undertake food-foraging tasks starts by positioning herself near to the hive entrance so that she can receive nectar from bees that are already foraging. Provided there is plenty of room in the hive in which to store the nectar, the bee will be keen and willing to take it, but there will be a reluctance to accept more if storage space is diminishing, especially if the nectar is of a lower quality than that offered by other foragers.

At about four weeks of age the worker begins her duties as a forager (collector of food) or a scout (finder of food), seeking out sources of pollen, nectar and water. It is unusual for bees always to find abundant sources of food close to their hive, so they must journey to wherever it is to be found. The food requirement of the bees changes during the season, and of course not all flowers are in bloom all season; indeed, some plants are not even in flower for the whole of the day.

Similarly, not all bees are occupied collecting the same thing: some may collect only pollen, others only nectar, while some collect both, and so on. Therefore the finding and foraging of food is a complex and well-controlled activity. When a scout bee finds a source of food she investigates it by landing on the flower, extending her proboscis and sucking up some nectar. Then she takes a bearing on the source of the food and notes the position of the sun (which bees can still discern

OPPOSITE: This bee is collecting drops of nectar which have exuded from the plant.

RIGHT: Workers going about their hive duties.

even if it is obscured by cloud) before returning to the hive.

Passing On the News

Once back in the hive, the scout will perform one of several types of movements, known as dances, that signal to other hive members the location and abundance of the new food source she has found. Forager bees will perform similar dances. Dances are performed in the darkness of the hive on the vertical surfaces of the comb. If the food source is close by (within about 80ft/25m of the hive) the bee performs a 'round dance', moving in circles while frequently changing direction. The greater the value of the food source, the more often the changes of direction occur. If the food source is further away, between 80 and 330ft (25–100m), then the bee will perform a 'waggle dance', in which manoeuvre she will move in a slightly flattened figure-of-eight pattern while waggling her abdomen from side to side in the

straight line she takes as she moves from one side of the figure-of-eight to the other. The distance of the food source is indicated by the duration of the straight run and the frequency of the abdomen waggles. These are accompanied by high-frequency buzzing sounds. Together, these actions may inform other hive members as to the quality of the food source. The angle at which the bee moves in the straight line

ABOVE & OPPOSITE: Worker bees dancing.

RIGHT: A worker bee will travel some distance from the hive to find food.

corresponds to the angle between the direction of the food source and the sun, when viewed from the hive entrance. The dances are performed to the accompaniment of antennae-touching and vibrations, all of which help to impart this vital information to other foragers.

FORAGING

A bee setting out on a foraging expedition to collect pollen, nectar, and perhaps water for the hive, faces a potentially hazardous journey. As if the task of going to and fro from the hive, laden with food, was not difficult enough, the foraging worker bee must

avoid all manner of natural and man-made hazards. Despite their stings, many animals see bees as the perfect meal: crab spiders lie in wait inside flowers for unsuspecting insect visitors, their bodies perfectly camouflaged to avoid detection until it is too late; bees fall victim to praying mantis and other

predatory insects; many birds, too, find bees delicious (there is no need to guess how the carmine bee-eater gets its name); while other animal enemies include frogs, toads and shrews.

Farming practices have also made life difficult for bees. A reduction in suitable crops nearby often means having to forage further afield, with all the dangers that entails, and crops or even garden flowers sprayed with insecticides are, of course, lethal to bees, not only to the foragers, in this instance, but also to other hive members that later make contact with contaminated bees or pollen. Man-made structures, such as moving traffic, also take their toll, as do houses and other buildings, where unsuspecting bees may become trapped and needlessly killed.

ABOVE & OPPOSITE: Worker bees out on foraging expeditions for food.

The state of the weather also has a major bearing on the success or otherwise of a bee's foraging activity. Persistent very dry weather or sudden cold snaps are not conducive to good foraging, and strong winds or sudden downpours of heavy rain can make flying difficult, particularly as a heavily laden bee is not the most aerodynamic of flyers.

THE DRONE

To the average person, drones are perhaps the least understood members of the hive. They are also a transient part of the community, produced in

OPPOSITE: The tongue of the bee opposite can be seen as it extracts the nectar from the flower.

RIGHT: The male honeybee drone serves little purpose other than to mate with the queen.

fairly small numbers for one task only at a certain time of the year. Drones are male bees, and they differ from the females of the hive (the workers and the queen) in a variety of ways. To begin with, they do not have ovipositors and thus they cannot sting, so they play no role in the defence of the hive. Nor are they involved in pollen or nectar collection since they do not possess the apparatus for this function. In fact, practically their only role is to mate with and fertilize new queens, although they may participate in helping to regulate the temperature of the hive. Drones have another important distinction: they are the product of an unfertilized egg laid by the queen, whereas workers come from fertilized eggs. Thus drones are the offspring only of the queen, and have single, unpaired chromosomes leading to their being described as 'flying gametes'. In temperate regions the drones that did not die after mating with the new queens are usually driven from the hive before the onset of winter and, because they cannot forage for themselves, they soon die of cold or starvation.

The drone is about 0.6–0.7in (15–18mm) in length or slightly smaller than the queen. He has a fuzzy, broad, blunt-ended abdomen, whereas that of the queen is long and tapered. Drones are easily recognizable by their compound eyes, which are large, extend to the front of the head, and meet in the middle. The large eyes help the drone to seek out queens on mating flights. It is also possible that they also help drones to spot predators, since they have no stinging defence mechanism of their

THE LIFE OF THE HONEYBEE

own. The drone has wings that extend the length of the abdomen, whereas queens have shorter wings.

Like other bees, a drone begins life as an egg laid in a cell. Usually, drone cells are located at the edges of the hive frame where temperatures are slightly lower. After hatching from the egg after three days, the drone spends six or seven days as a larva before pupating for about two weeks, and then emerges as an adult (a total of about 24 days). Workers feed the developing drone. It receives a fairly nutritious diet, richer than a worker's but not as rich as a

OPPOSITE: A postnatal image of an *Apis mellifera carnica* drone.

ABOVE: The drone's very large eyes enable it quickly to identify queens, out on their mating flights.

queen's. Workers also feed the emerged adult drone, even though he can feed himself on honey stocks stored in the hive. Like the queen, the drone learns the lay of the land around the hive by embarking on a series of orienteering flights. In due course the drone sets off on its mating flights to fertilize the queens. Drones from one hive do not mate with queens from the same hive, and this helps to prevent inbreeding and keeps the gene pool healthy.

CHAPTER SIX
BEE PRODUCTS

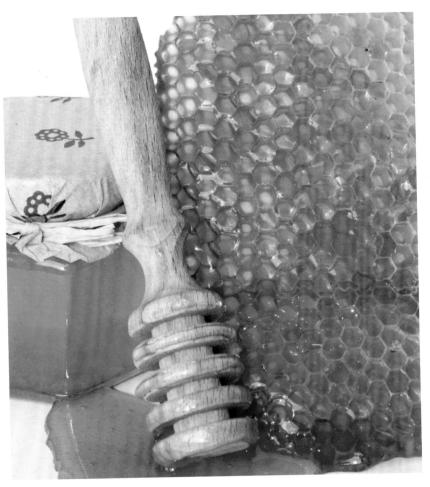

For the bee-keeper, the prime fruit of a honeybee's labours is, of course, honey, but bees collect or produce several other products as well. In a later chapter we shall look in detail at the importance not only of honey but also at some of the other bee products that are of value to mankind.

BEES & BEE-KEEPING

HONEY

Bees create honey as a food source for themselves, using it particularly when other food is scarce or the weather is cold. Honey is derived from the nectar of flowers. Its sweetness is well-known, and comes from the glucose (about 31 per cent) and fructose (about 38 per cent) sugars that are formed by the action of the bee's enzymes working on the nectar. Honey also contains about 18 per cent water, small amounts of another sugar, known as sucrose, and various minerals, vitamins and enzymes.

Honey, with its golden goodness, is an evocation of long summer days, the heady scent of many flowers, and the contented drone of the bees as, seemingly spoiled for choice, they flit from plant to plant.

BEE PRODUCTS

Once a foraging bee has returned to the hive with the sugar-rich nectar in its honey sac, the nectar is regurgitated several times until the partly digested substance reaches the desired stage and quality and is ready to be stored in honeycomb cells. The water content of the honey is reduced by the action of the bees fanning their wings to increase evaporation. Then the cell is capped with a waxy lid until it is required. Due to the unique antibacterial and antifungal properties of honey it has a long 'shelf life' and will not ferment or spoil provided it is properly sealed. There is much more to tell about honey in later chapters of this book, including how to harvest it and its many uses, from the purely medicinal to its role as a key ingredient in many drinks and foods.

ABOVE: Pure honeycomb straight from the hive. Fresh, new comb is sometimes sold and used intact as comb honey.

OPPOSITE: Honey, an elixir of life?

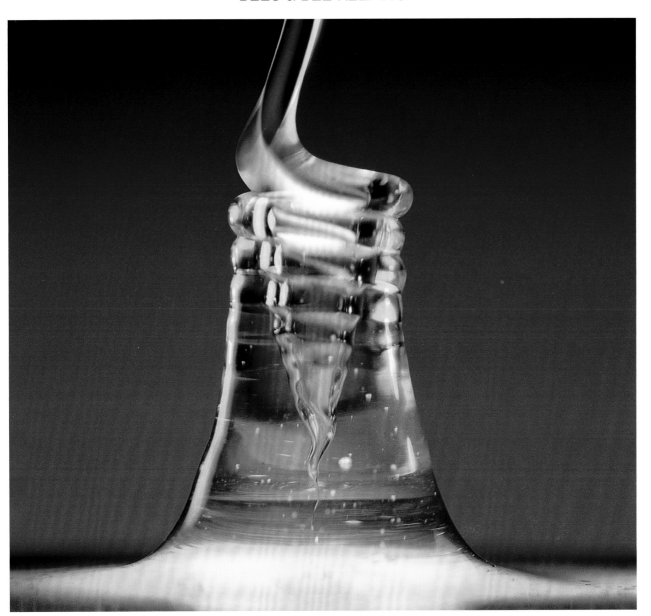

BEE PRODUCTS

bloom, and this results in a well-balanced mix of nutrients. Bees get protein, starch, fat, minerals and vitamins from pollen, which they need to build up their bodies, feed their young, produce wax, and make their venom for defending the hive. Pollen is vital as a food for developing larvae and young adults as well as for producing brood food (royal jelly). If, for any reason, pollen becomes scarce, then the whole colony output will be compromised as a result.

When a foraging bee returns to the hive laden down with her cargo of pollen attached to the pollen baskets on her legs in the form of small pellets, she immediately places it into a cell which may already contain some pollen. Other bees use their heads to pack the pollen in tightly in order to conserve space. Pollen is stored near to the broodnest, where it is readily available for use. At the same time, a little honey, nectar and some glandular secretions are also added by the worker bees to help increase its nutritional value.

POLLEN
Pollen, taken from the anthers of flowers, is the main source of food required by bees apart from carbohydrates and water. In the course of a season, a colony of bees will collect pollen from a wide variety of different flowers as they come into

ABOVE LEFT: A worker bee collecting pollen.

OPPOSITE: The pollen is then stored in baskets on the back legs to transport back to the hive.

BEESWAX

Beeswax is a natural form of wax produced by the Apis species of bees by the time they are about 12 days old. Worker honeybees have wax glands on the fourth, fifth, sixth and seventh abdominal segments. The wax secreted from the glands is first manipulated by the bee's legs and then by her jaws or mandibles. The wax is used in the building of the walls of the honeycomb cells and to cap or seal them off when they have been filled with honey. Wax cells are initially clear and colourless but become white after having been chewed by the worker bee. By the time pollen and propolis are added, the wax turns yellow or brown. For additional strength, the wax may be supplemented

OPPOSITE LEFT: Beeswax straight from the hive can be quite dark in colour.

OPPOSITE RIGHT: Beeswax is used to make candles that produce a natural and attractive fragrance as they burn.

RIGHT: Bees use wax to cap the honeycomb cells to seal in honey as well as larvae.

with propolis. As we shall see later, bee-keepers also harvest wax, along with honey, for which there is a variety of uses.

ROYAL JELLY

Mention has already been made of this vital substance, produced by young worker bees in their hypopharyngeal glands. It is one of the substances that is fed to all larvae for a time and is instrumental in the development of queen bees. Later, we shall look at how the production of royal jelly is encouraged by bee-keepers, how it is harvested, and its uses as a dietary supplement.

PROPOLIS

Propolis is a sort of general-purpose building agent that honeybees make from the resins that plants produce to protect their buds against invasion by bacteria and fungi or to prevent insects from eating them. Plants used to make propolis include poplars and various species of coniferous trees, as well as a few types of flowers. When the bee is out foraging, she will scrape a tiny amount of the resin from the plant with her jaws (the plant is unharmed by this process), transferring it to her back legs for transportation back to the hive. On her return, the resin is mixed with enzymes from the bees to make it workable and it is also often mixed with a little pollen.

Propolis is then used by the bees to seal small spaces in the hive to help keep out the elements or unwanted animal visitors; larger spaces are first usually filled with beeswax. The same antibacterial and antifungal properties of the resin that protect the plant from which it came are also used by the bees to keep these agents out of the hive. Occasionally, a small animal, such as a shrew, may enter a bee's nest and die.

BEE PRODUCTS

LEFT: A build-up of propolis in the hive.

BELOW: Propolis is a resin bees take from plants. It is used to seal the hive, protecting it from the elements and unwanted visitors.

OPPOSITE: A swarm occurs when the hive becomes too full. Then, the queen, accompanied by workers and a few drones, will depart en masse to form an new colony, leaving a new queen to carry on in the old hive.

warm weather if they are glued together with propolis. But the same part of the hive is likely to open with a snap in cold weather as the propolis seal is broken.

Since the bees cannot remove such a large object, they instead encase it in propolis, relying on the substance's properties to effectively 'mummify' the body and prevent it from releasing harmful bacteria into the nest.

The colour of propolis varies according to the plants from which the resin was obtained, but it is usually brown, although it may range from grey to almost black. At room temperature propolis is sticky and has a consistency resembling chewing gum. When it is cold, however, propolis becomes hard and brittle. Bee-keepers often find it difficult to open parts of the hive in

Propolis is sometimes sold as a health supplement and as a traditional medicine, though claims for its effectiveness in some treatments are still being evaluated.

SWARMING

Once a colony has become too big (in other words, when all the available space has been filled with eggs, larvae and food) and is about to swarm, there are various signs of activity that a bee-keeper will notice. The workers will begin to build queen brood cells; these are larger than ordinary ones and are built so that they hang down from the bottom of the hive frame. Also about this time, although perhaps not so readily discernible, is the fact that some of the foragers are turning their attention away from food-gathering and have begun prospecting likely new sites for a home instead. When the first of the newly-laid queen larvae have reached the point of pupation, and assuming that the weather is favourable, the old queen, about half of the workers in the hive and several drones, will suddenly leave en masse. The rest of the colony inmates will remain behind to continue life as normal in the hive, along with the

newly emerging queen. Soon she will undertake her mating flights and begin, what will be for her, the almost endless task of egg-laying.

Once the swarm of departing bees has left the hive, the bees swirl around outside in close proximity to the hive before heading for a gathering place, which is often on the branches of a tree about 150- to 200-ft (46–61-m) tall. Then scout bees inform workers in the

swarm of the location of possible future nest sites, indicating their positions by their dances. Finally, a site is selected and the rest of the swarm move to take up residence. The first task is to set about building a new comb so that they can raise new young and store their food. Foragers begin the task of bringing food and water to the hive, the queen begins laying eggs, and soon the new colony is in full swing.

CHAPTER SEVEN

BEES, HONEY & MAN

Mankind's association with the bee goes back to earliest times, the link, no doubt, first having been formed, and then strengthened, by the discovery that the honey produced by bees was a valuable food for humans as well. In time it was also seen as a symbol for sweetness itself, and the word 'honey' is commonly used as a term of endearment today. It is

possible that early man came across a nest that had already been raided by an animal, such as a bear, or had been exposed when a tree housing the nest had been struck by lightning or damaged by wind. It is believed that humans first started to hunt for honey about 10,000 years ago. Cave art found near Valencia in Spain, and believed to be 8,000 years old, shows a Mesolithic rock painting of a female collecting honey

and honeycombs from a wild bee's nest inside a tree. The woman is using a kind of ladder to reach the nest and there is a basket for carrying away the spoils. The artist does not record how many stings the woman may have received for her trouble!

In time, humans discovered, again perhaps by chance, that smoke from a fire acted as a calming influence on bees (see also page 167 et seq.), following which depictions of honey collection are often seen showing gatherers armed with smoking torches. The walls of the sun temple of Nyuserre, from before 2422 BC, show workers blowing smoke

FAR LEFT & OPPOSITE: Honey is one of the most extraordinary and ancient foods still in existence, valued from the time that Stone Age hunters first raided wild honeybee nests.

ABOVE LEFT: Then it was discovered that smoke from a fire calmed bees. This method is still used today, albeit using a more modern device.

BEES, HONEY & MAN

into hives as they remove the honeycombs.

Reflecting the importance of honey to the people of ancient Egypt,

the honeybee was chosen as the symbol of Upper Egypt following the unification of the north and south, and the same inscription continued to be used from the 3rd Dynasty (about 2650 BC). The ancient Egyptians used honey to sweeten cakes and biscuits and in many other dishes, and because of its preservative properties it was also used in the process of embalming the dead. Depictions on the tomb of Pabasa from the 26th Dynasty (about 650 BC) show cylindrical beehives and people pouring honey into jars. The ancient Greeks also had a fondness for honey, and there is evidence that a well-developed honey industry existed in the Holy Land in biblical times, around 3,000 years ago. Archeologists examining the ruins of the city of Rehov found over

100 hives, some still intact, made from straw and clay. This is one of the very earliest apiaries ever discovered, even though the practice of bee-keeping itself predates these ruins.

In his book, *Naturalis Historia*, published between AD 77 and 79, the writer, naturalist and philosopher Pliny the Elder devotes much space to the bee and honey, as well as to the many uses to which honey may be put. In parts of Byzantine Greece, such as Rhodes, it was once the custom for a new bride to dip her fingers into honey and make the sign of a cross before entering her new home, while in the Roman Empire, honey was sometimes used instead of gold for paying taxes.

Although honey continued to be an important culinary ingredient

BEES & BEE-KEEPING

throughout the ages, information on bee-keeping in medieval times is somewhat scarce. We do know, however, that abbeys and monasteries in this period were especially important centres of bee-keeping. Not only was honey an important food and sweetener, but fermented honey was also used to make the alcoholic beverage known as mead. Mead itself

LEFT: The symbol for Manchester in England is a worker bee, seen in places all over the city.

BELOW LEFT: A bee relief in the Vatican, Rome.

BELOW: Roman beehives on Malta.

has a long history of production; it is older than wine and was well-known to the ancient Cretans. In addition,

beeswax became vital in the making of candles – the most practical form of lighting in those times, and indeed for many centuries afterwards – as well as for making wax seals to ensure the privacy of letters and documents before the invention of glued envelopes. Important dignitaries, such as church officials and royalty, would also 'sign' or endorse documents by pressing special rings into wax seals before they set hard, thus leaving lasting impressions.

Pictures exist in manuscripts from medieval times of wicker hives and woven straw baskets known as skeps. There are also written accounts of the removal of bees to heather moors, so that they could visit the flowers; on feeding bees in winter; and on the use of sulphur to kill bees in order to collect their honey and wax.

At some point in history, humans began to keep wild bees in artificial nests, or hives, constructed from hollowed-out logs, wooden boxes, clay

vessels and skeps. Many of these structures needed to be destroyed before the honey and other contents could be harvested, which resulted in the wasteful loss of entire colonies. But the 18th and 19th centuries saw

ABOVE LEFT & ABOVE: The time came when wild bees were eventually caught and kept in man-made nests, including woven straw skeps.

OPPOSITE: This traditional Portuguese beehive with a pantile roof is made of cork.

progressively more advanced techniques in bee-keeping and hive construction,

OPPOSITE: Victorian beehives in the Lost Gardens of Heligan, Cornwall, England.

ABOVE: A novelty beehive in the shape of a house or church.

which meant that colonies could be preserved once the honey and wax had been harvested. In the 19th century the evolution of the modern hive resulted in the movable comb hive, designed by Lorenzo Langstroth, whose name is still well-known among bee-keepers today, since the Langstroth hive is one

of the most popular types available, especially in the United States.

Today, although the design of beehives has advanced from the earliest types, the art of bee-keeping is still much as it has been for hundreds of years. And in some cultures, in parts of Australia, Africa, Asia and South

America, the method used for collecting honey from the wild by breaking open nests still more or less prevails as it did when humans first began raiding bees' nests thousands of years ago.

In parts of Africa and Asia, one or two of the species of birds, known as honeyguides or indicator birds, use remarkably cooperative behaviour to help them get their food. Honeyguides

Beehives come in all shapes and sizes.

feed on beeswax and the larvae of wax moths (waxworms) that frequently inhabit bees' nests. They get their name from the fact that they guide humans and other animals, such as ratels (honey badgers), to bees' nests. Once the honeyguide spots the nest (which it is unable to open by itself), it flies off to look for an 'ally'. The honeyguide's excited behaviour, which includes flying for short distances, then waiting, leads the ally to the nest. Once the human, or the ratel, opens up the nest to steal the honey, the honeyguide swoops in to feed on the exposed beeswax and larvae.

BEES AND HONEY IN MYTH AND
RELIGION

Bees and honey have long featured in
religion. Alongside beehives discovered
at Rehov, an important Bronze and
Iron Age Canaanite city built upon a
large earthen mound in the Jordan
Valley in Israel, an altar was discovered
decorated with fertility figurines which

OPPOSITE: This colourful beehive resembles a
Swiss chalet.

LEFT: The Langstroth is the most commonly
used beehive in America.

ABOVE: Lorenzo Langstroth, inventor of the
famous beehive.

may indicate ceremonies associated with bee-keeping. In the Jewish tradition, honey is also used symbolically at Rosh Hashanah, when apple slices are dipped in honey to secure a good and sweet New Year. In ancient Near Eastern and Aegean cultures the bee was regarded as a sacred insect that formed a link

between the world of the living and the underworld or that of the dead. Mycenaean *tholos* tombs were shaped like beehives, while the Homeric Hymn to Apollo tells how the god's gift of prophecy was given to him by three maidens, usually identified with the Thriae, these being a trinity of pre-Hellenic Aegean bee goddesses. Images of the Thriae appear on embossed gold plaques recovered at Camiros in Rhodes, and which date from the 7th century BC, although worship of the goddesses is considerably older.

Honey figured in the diet in most Mesoamerican cultures, besides being an important trade product, and Ah-Muzen-Cab, the 'Great Bee God', was a deity of the Mayan pantheon. The Mayan word for 'honey' was also the same as that for 'world', indicating that the god was a creator of the world. The San people of the Kalahari Desert in Africa tell a story that is their particular version of the creation myth in which a bee is transporting a mantis (another kind of insect) across a river. The exhausted bee leaves the mantis floating on a flower but not before it has planted a seed in the mantis's body. The seed grows to become the first human being.

There are many references to honey in the Bible's Old Testament. In the Book of Judges, Samson, on the way to be married, sees the carcass of a lion that he had previously killed with his bare hands. He notices that bees have nested in the carcass and are

ABOVE LEFT: Wahlberg's honeyguide (*Prodotiscus regulus*), an extraordinary little bird which feeds regularly on beeswax, thereby indicating the presence of wild bees' nests.

FAR LEFT: It is a Jewish tradition to dip apple slices into honey at Rosh Hashanah.

ABOVE: Ah-Muzen-Cab, 'The Great Bee God' of Mayan mythology.

producing honey. At the feast, Samson sets a riddle which will be rewarded with fine linen and garments if his 30 groomsmen can solve it. The riddle, 'Out of the eater, something to eat; out of the strong, something sweet' is therefore based on Samson's second encounter with the lion. In the Book of Exodus, the Promised Land is described as 'a land flowing with milk and honey', while in the New Testament (Matthew) John the Baptist is said to have survived in the wilderness for a lengthy period on a diet of locusts and honey.

As recounted by the historian, Josephus, the name of the Hebrew prophet and leader, Deborah, meant 'bee'; she was known for her industry, wisdom and sweetness of temper towards her friends, but for her sharpness towards her enemies. Honey plays an important role in the Buddhist festival of Madhu Purnima, which is celebrated in India and Bangladesh, and which recalls Buddha making peace among his disciples and entering the wilderness. While there, legend has it that a monkey offered him honey to eat, and Buddhists remember this act of kindness by giving honey to the monks at Madhu Purnima.

CHAPTER EIGHT
KEEPING BEES

Although bee-keeping is a long-established commercial operation in many countries of the world, creating a source of employment for many as well as providing honey and other important by-products, it is nevertheless an activity that can be pursued by interested individuals in a small way, as little more than an interesting hobby, while at the same time producing some honey for personal use or for giving away to friends. It can also be a sociable pastime, for bee-keeping associations exist not only to provide information and advice, but also to be places where like-minded people can meet.

It is hardly surprising that keen gardeners should often be just as interested in bee-keeping, for it is a good way of ensuring bees visit their plants and pollinate them. To others, of course, honey is the key reason for keeping bees, and a well-managed colony, in a suitably flower-rich

LEFT: Traditional skeps made from woven straw.

OPPOSITE: The Museum of Ancient Bee-keeping in Stripeikiai is the only one of its kind in Lithuania, established in 1984 by Bronius Kazlas and his wife. The wooden sculptures also illustrate the importance of the bee in the mythology of different cultures: Egyptian, Native American and Lithuanian.

KEEPING BEES

BEES & BEE-KEEPING

Bee-keeping on a large scale at the foot of the Great Smoky Mountains in Tennessee. The traditional sweeteners of the area include honey, sorghum and maple sugars. Initially, honey would have been gathered from wild beehives, but farmers began domesticating bees, keeping them in hollow tree hives, then box hives. Popular in the area is sourwood honey, the sourwood trees growing to a height of 3–5,000ft (915–1525m), reaching their maximum on the western slopes of the Great Smoky Mountains.

environment, might expect to yield about 50lbs (22.5kg) of honey a season, and possibly quite a bit more. Keeping bees for their honey is rewarding enough, but it is a rare privilege to observe one of nature's most enduring and fascinating creatures at first hand.

THE RIGHT CHOICE?
Bee-keeping won't suit everyone: although keeping bees does not require the sort of stamina or time needed to look after, say, horses or some other livestock, it nevertheless calls for certain things to be done at certain times. Emergencies or potential problems must be dealt with promptly, taking the right course of action, and the whole enterprise needs to be set up

and maintained correctly to ensure a happy and successful hive. Remember, you are working with live animals that to some extent will depend on you for their welfare. Another obvious factor to consider is the bees themselves. The presence of a beehive or two in your garden is going to increase significantly the number of bees buzzing around the area, especially in the vicinity of the hive itself, and this may be an issue with certain family members. Then there is the fact that bees sting, and as a bee-keeper it is almost inevitable that you will get stung several times in a season, the chances of which can be much reduced by good bee and hive management.

It would be wise to consult any neighbours whose properties are adjacent to the intended site of your hive or hives, just in case they have valid objections – such as strong allergies to bee stings. It is also advisable to check that there are no covenants, local bylaws or other restrictions preventing you from keeping bees. Remember, too, that although bee-keepers tend bees and

As far as the continuation of plant species is concerned, bees are as vital in Ethiopia as they are in an English country garden.

provide them with a home and even nourishment when required, these insects will never become tame or domesticated like some other 'farmed' animals. The bee-keeper needs to work with the bees, interpreting their needs and moods and reacting accordingly, while at the same time handling them with gentle firmness and confidence.

Bees live and work to a seasonal cycle, geared closely to the availability of flowers and the state of the weather. Therefore the time you spend attending to the hive will also vary. As a rule of thumb, half an hour to an hour per

week needs to be spent on a colony during the summer months, but this may well reduce to fortnightly inspections as your experience grows. In winter, in temperate climates, when foraging ceases and general activity is very much reduced, most of your work is likely to be concerned with maintaining your bee-keeping equipment. But remember that there is honey to harvest and process, not to mention gathering and using some of the beeswax, all of which take time. But then, wasn't this all part of the attraction in the first place?

MAKING A START

Before going to the trouble and expense of getting the bees for your hive and buying all the necessary equipment – and indeed, there is plenty of it on offer to tempt the unwary – it is worth attending demonstrations, such as those often held by local bee-keeping associations. Here you will see exactly what is involved in opening the hive and handling the bees, and there will, of course, be opportunities to ask knowledgeable and experienced bee-keepers to answer the questions that have been bothering you.

It may even be possible to handle bees yourself so that you can experience exactly what to expect at first hand. This may well be the moment when you discover that the prospect of potentially venomous little creatures, wandering all over your body (even if you are wearing protective clothing), is not to your liking, and that bee-keeping is not for you after all. But if you relish the thought and decide to proceed, make a

RIGHT & OPPOSITE: The idea of keeping bees is an appealing one. Be wary, however, of causing annoyance to neighbours, and be prepared to look after your bees properly.

note of the people present who may best advise you later on. Bee-keepers often have stands at country shows and other similar events, and although actual bees may not be present, there will probably be leaflets and other useful information, and someone expert enough to discuss the subject with you. It will also be of benefit to join one of the bee-keeping organizations in your area, or enrol in a starter course in bee-keeping, perhaps through a local bee-keeping club, which is a valuable way to approach the subject in the correct manner.

There are plenty of websites and books on the subject of bee-keeping, and it is worth looking at as many as you can in order to study the subject from all angles, while magazines, leaflets and other periodicals may give other points of view. Don't forget, however, that not all of these sources of information necessarily relate to, or

OPPOSITE: Should you decide to keep bees in your backyard, remember that far more bees than usual will be buzzing about in the area.

RIGHT: Producing your own honey is immensely satisfying as well as providing an excellent food for family and friends.

originated in, the country in which you live or intend to keep bees, so be aware that some specific topics or discussions concerning equipment and techniques may not apply everywhere. From time to time you will come across articles written by amateurs and others new to bee-keeping. It can be quite informative – and sometimes amusing, too – to read about the experiences of others, particularly since they may forewarn you of the pitfalls into which you could easily fall yourself.

It is necessary to consider the best place in which to position a hive before you purchase any equipment, and then set about preparing the site in anticipation of its arrival. It goes without saying that the hive should be placed on firm, dry, weed-free ground. A well-ventilated sunny aspect is also ideal, perhaps one that faces south or east. Locate the hive away from danger of frost, dripping water and wind if possible. Don't forget that you will need

LEFT & OPPOSITE: Before embarking on your own bee-keeping venture, it is important to seek the best expert advice available and even to attend lessons, when the instructor will demonstrate the correct way to handle bees and also advise on the best equipment to buy.

comfortable access to the hive yourself, while a hive stand will make the task that much easier; some hives come with legs already attached and these may give you all the height you need. The stand must be tall enough to keep the hive out of reach of marauding animals and strong enough to support the weight of a healthy colony, which may eventually attain about 150lbs (70kg) or so. You can also make a hive stand out of breeze blocks, bricks or stout timber, which should be about 2–3-ft (0.6–0.9-m) high. This allows access to

ABOVE: Children are certain to be fascinated by bees, but strict supervision is vital at all times.

OPPOSITE: Bees can be kept almost anywhere; in fact, this beehive in sited on the roof of a Chicago penthouse.

the hive without your having to bend too much and will keep the hive itself clear of damp ground. Whatever system you use, make sure that the hive is secure on its stand and cannot

wobble or, worse still, blow over in a strong wind. Don't forget that you may need to access the hive from all sides, so position it with plenty of space surrounding it. You will also need to

OPPOSITE & BELOW: Whatever the location, keeping bees can be successful as well as of benefit to the environment.

ensure that the hive entrance doesn't become obscured by weeds and other plant growth.

Fresh water must be available at all times; if there is no natural source of water close by, then one must be provided. This can be a moderately large, shallow dish sited off the ground, a shallow-sided bird bath being ideal for the purpose; but whatever is used must allow the bees access to the water without the danger of them falling in and drowning. If possible, screen your hive to prevent it from becoming an eyesore for neighbours, and remember that the bees' flightpath should be as far as possible from places where you or your neighbours are likely to congregate; no one wants squadrons of foraging bees flying across their patios while they are enjoying their gardens.

The flightpath should also avoid roads, and a tall fence or some natural plant cover, such as evergreens planted close to the hive, will ensure the bees' flight takes them high and away from hazards when approaching and leaving the hive, thus minimizing disturbance. And finally, it goes without saying that if it wasn't already, your garden should now become an insecticide-free zone, for insecticide-laden sprays can also be

blown towards hives by the wind as well as covering adjacent plants.

BASIC EQUIPMENT

There is equipment galore available from a myriad of suppliers, although much of it will prove unnecessary for the beginner or even for the experienced small-scale bee-keeper. But once you have decided what you do need, the best advice is to buy the best you can afford and then to look after it so that it will give you many years of

reliable service. It is certainly worth checking the prices of similar items before finally deciding to buy. Make sure you order or obtain all the equipment that is really necessary beforehand, so that everything is to hand once the hive is set up and running. For instance, it's no good wondering how to feed a nucleus of

OPPOSITE & ABOVE: Bee-keepers should buy specialist clothing for their own protection. An all-in-one beesuit is probably the best choice.

KEEPING BEES

hungry bees after they have been delivered. For this reason, let's start with the essential equipment before considering the hive itself.

PROTECTIVE CLOTHING
To reduce the chances of being stung and to keep your clothes clean while tending to the hive, either a beesuit or a bee tunic or jacket, available from bee-keeping suppliers, is an essential item. A beesuit is a one-piece unit intended to cover and protect the entire body and which resembles one of the spacesuits worn by astronauts. The drawbacks of beesuits are that they are cumbersome and can be uncomfortable to wear in hot weather. A bee tunic or jacket protects the head, upper body and arms, and should have elasticated sleeves. With this form of clothing, you will, of course, need to make provision for protecting your legs by wearing some loose-fitting strong trousers, that can be tucked snugly into your boots; in this respect, a pair of rubber

LEFT: The head should be thoroughly protected at all times.

OPPOSITE: Novices, in particular, should wear heavy-duty gloves.

OPPOSITE: The correct clothing does much to make bee-keeping a pleasurable occupation while keeping safe.

BELOW & PAGES 166–167: It is sometimes useful to work with another person, but remember that interested bystanders should also be well-protected.

'Wellington'-type boots are ideal, but make sure they are wide enough at the tops to take your trousers and provide a gap-free fit when they are tucked inside. Whatever style of clothing you have, it is best to choose a hood and veil (to protect your head and face) that can be detached by a zipper or similar secure method to prevent bees from getting inside. You can, of course, improvise by choosing a wide-brimmed hat with a fine mesh securely sewn to the brim all the way round. The mesh must be sufficiently long to be tucked in securely at the neck so that bees cannot enter at this point. Most proprietary protective clothing tends to be white, with a darker-coloured mesh

or veil for protecting the face. There is a reason for this; in the wild, the bees' natural hive-raiding enemies include bears, skunks and other predators. These tend to be brown or black in colour, so white or light-coloured clothing helps the bees to distinguish the bee-keeper from their usual predators.

Gloves are an essential item for the beginner, even though they can restrict movement and tend eventually to be discarded by more experienced keepers. Appropriate gloves are available from suppliers, but a pair of strong, close-fitting rubber gloves are a good substitute. Some stores sell strong but supple gloves for general maintenance and DIY, which may also be suitable, but the purpose-made gauntlets, that go some way up the wearer's sleeves, offer the greater protection.

SMOKERS

An essential item of equipment as far as the bee-keeper is concerned. There is quite a variation in quality and ease of use in respect of smokers, so try to look at and compare a few different types before making a final choice. A smoker is a device used to puff smoke into the colony with the purpose of calming the

bees so that the hive can be inspected more easily. It is shaped rather like a coffee pot, and it is essentially a metal can with a spout at the top and a grate at the bottom in which fuel is burned (to make the smoke); there is a simple bellows arrangement attached to it so that the smoke can be forced out of the spout. Choose a larger rather that smaller one, and be sure it has a guard around it to prevent the hot sides of the device from touching you or your clothing. Some smokers have quite stiff bellows, making them difficult to operate, so opt for one with a nice smooth action.

A variety of fuels can be burned in the smoker: ready-prepared fuels can be bought from bee-keeping equipment suppliers or you can use your own, or a mixture of both. Commercially available fuels include rolls of corrugated cardboard, cotton waste, hessian, old sacking and compressed pellets of wood chips. Your own fuels should consist of sawdust, dried and crumbled natural timber (for example, rotten wood), pine needles and so on. Old building timber, cut up and pulped, can also be used, but never use anything on which an insecticide or wood preservative has been used, since

the fumes from such wood can be toxic to both bees and human beings. For this reason, avoid using pieces of wood from decking and fence offcuts, since many of these have been commercially pre-treated using harmful chemicals. Whatever fuels you choose, always use them in accordance with the instructions supplied with the smoker, and check first with an experienced bee-keeper or someone from one of the bee associations if there is any doubt. It goes without saying that young children should be allowed nowhere near a working smoker.

So how does the smoke work? The calming effect of smoke on bees has been known since ancient times. First, smoke masks the pheromones that are the bees' chief means of mass communication and which are used by guard or injured bees to alert the hive to potential danger, such as when it is being opened. This breakdown of communication, and thus the chain of command and action, gives the bee-

LEFT & OPPOSITE: The smoker is an essential piece of equipment as far as the bee-keeper is concerned; the smoke calms the bees, making inspection of the hive that much easier.

keeper an opportunity to check the hive and close it again without too much disruption to the occupants. The smoke also seems to initiate a feeding response in the bees, presumably

OPPOSITE & ABOVE: Special fuel for smokers can be bought or you can use your own.

because the disruption causes them to prepare for abandonment of the hive; therefore, they are stoking up on food to be used when the colony first re-establishes itself somewhere safer. Moreover, a fully engorged bee is less likely to sting, so smoke has another positive effect in protecting the bee-

keeper. Using the smoker correctly is an acquired art, and the inexperienced bee-keeper should not be surprised or dismayed if the technique fails to work as it should at first. Small, well-directed puffs of smoke, rather than a massive smokescreen, are what you are trying to achieve.

HIVE TOOLS

The hive tool is a specially designed piece of equipment that is used as lever, scraper and hook. The tool comes made from wood or stainless steel, and is often brightly coloured since it seems to end up frequently getting lost. It is used when dismantling the hive and when removing frames from it, the scraper part being utilized to remove build-ups of propolis from parts of the hive. Such a range of functions would suggest a complex tool, but in fact the design is quite simple. One of the best types (about 10in /25cm) resembles an elongated version of the kind of scraper used to remove ice from car windows in winter. One end is flat, broad and sharpened and acts as a scraper, and the other is curved, with a slot near the end, and is used as a lever. When using the tool to separate parts of the hive, the curved end of the tool is placed between adjacent frame-top bars and twisted to left or right. Levering is achieved by lipping the curved part under the item to be lifted and then pressing down on the other

The calming effects of smoke on bees has been known for centuries, and when used properly is an effective way to approach them safely.

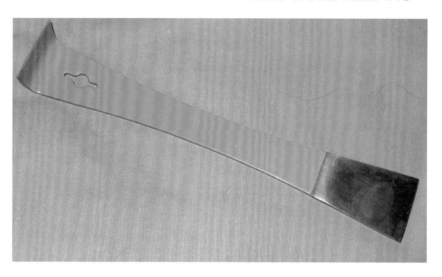

LEFT & BELOW: The hive tool has a number of uses and is a vital piece of equipment as far as the bee-keeper is concerned.

You will soon see that several different types are available, both new and secondhand, some of which are offered ready-assembled, with others coming 'flatpacked' for home assembly. The 'Langstroth' is used almost exclusively in the United States, but also appears in Britain and other parts of the world, while popular hives in Britain are marketed under names such as 'WBC',

end. The hole in the tool is intended for removing nails, but is also a convenient place to attach a cord so that the tool can be tied to your belt, making it conveniently to hand when required.

HIVES

This is clearly an extremely important item of a bee-keeper's equipment, for it is the place where the honeybees will live and from which honey and other bee products will be harvested. Like so much else in life, beginners can easily be confused by the available choice and may ultimately be seduced by an attractive design, a keen price, or even a supplier's sales patter, and thus buy the wrong type of hive for their purpose.

BEES & BEE-KEEPING

These beehives have been sited in a field of lavender both to promote pollination and to produce a unique honey with a wonderful flavour and aroma.

'Commercial' and 'National'. The most popular hives are made of wood, though artificial materials, such as polystyrene, are also used. Cedarwood is a popular choice in temperate climates, since it is very durable yet light in weight. Secondhand hives carry the risk of being infected with foulbrood, a disease of honeybees caused by the bacterium *Melissococcus plutonius* or by the more virulent *Paenibacillus larvae* (see Chapter 9). If there is any doubt concerning the provenance of such a hive then it is advisable to steer clear of it altogether.

Hives are constructed as a series of boxes, each carrying suspended frames on which the bees build their combs. Above these brood-frame boxes there can normally be fitted a meshlike device known as a queen excluder, the mesh allowing the workers to pass

LEFT: The hive tool should always be to hand.

OPPOSITE: Cross-sections of Langstroth, National and WBC beehives.

PAGE 178: A bee-keeper working on his Langstroth hives.

PAGE 179: A colourfully painted WBC hive.

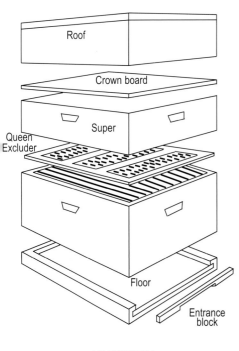

LANGSTROTH

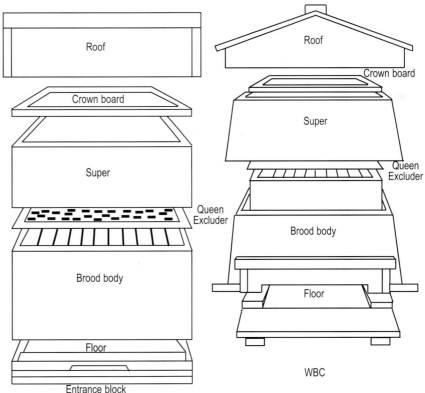

WBC

NATIONAL

through freely while restricting the queen because of her larger size. Because she cannot access any frames above the queen excluder, she cannot lay eggs in them, and they are therefore used only for storing the honey. Frames or boxes used for storing honey are called 'supers', and they are obviously of great interest to the honey-collecting bee-keeper. By altering the position of the queen excluder, a bee-keeper can decide how many boxes to allocate to the brood section or to the super section of the hive. (In the active part of the season, when there are plenty of flowers from which to forage, many bee-keepers consider it worth having two supers in use at the same time.)

Each frame consists of a four-sided structure somewhat like a picture frame. Frames (left) may be made of wood or plastic, within which there is either a beeswax-covered plastic foundation sheet embossed with a

honeycomb pattern, or a sheet of honeycomb-embossed pure beeswax set on a wire frame. In either case, the bees build their comb on the foundation sheet. It is advisable not to mix frames in the same box that have been used by bees for brood-rearing and food storage.

At the top of the uppermost box there should be an inner cover. This allows for a gap between the outer top covering of the hive and the hive itself, and helps the airflow as well as preventing heat from the outer cover acting directly on the hive in hot weather. Also, without the inner cover,

there is a tendency for the bees to seal the outer cover to the top frame with propolis, making its removal difficult.

OPPOSITE: National beehives.

ABOVE: Attractive WBC hives sited at the bottom of a country garden.

Even the inner cover will need to be removed with the flat blade of a hive tool, but this is an easier prospect than levering under the outer cover. To avoid the need to ease the inner cover off, some bee-keepers lay a sheet of canvas over the top frame instead, which can be rolled back when it is necessary to inspect the bees.

Another device is a clearer or an escape board. This is designed to remove the bees from the supers down to the lower boxes and prevent them from returning. This allows the supers to be removed for honey extraction.

The outer cover is a durable waterproof structure at the top of the hive boxes that keeps the weather out. It is usually constructed from wood or plastic. At the base of the hive there is a bottom board with a place for the bees to enter and leave, and usually a platform or ramp that acts as a landing stage and waiting area so that incoming bees can be inspected by the guards.

OPPOSITE: Hives sited in an apple orchard to ensure pollination and the production of fruit.

RIGHT: Before the onset of winter, hives may be placed in a more sheltered position to protect them from cold winds.

183

Some bottom boards are fitted with screens to allow debris to fall clear of the hive and to increase ventilation within.

A competent DIY enthusiast can easily make a serviceable hive, but it is recommended that good-quality plans and dimensions are followed rather than merely copying a diagram from a book since, although the overall size of the hive is somewhat unimportant, some of the internal dimensions, spacings and structural details are critical if the hive is to be fully effective. For example, there must be the correct space at the tops of the frames so that the bees can walk about, but not so much that they start building combs there. Take heed of advice concerning the treatment of the surfaces, so that nothing toxic is used that could harm the bees. Suppliers also sell specially designed and safe weather coatings for hive exteriors.

Many authorities recommend beginners start with a hive consisting of three or four medium-depth frames, but once some experience has been gained, and there is an expanding colony, more can be added.

LEFT & OPPOSITE: Bees coming and going as they please.

FEEDERS

Feeders are devices that dispense food to bees in the form of sugar syrup. They come in a variety of forms: hive-top feeders are designed to replace the inner cover on the top of the hive, and are made of wood or plastic. They hold up to about 2 gallons (9 litres) of sugar syrup, and can be used at the end of the year to feed the colony. The design of the feeder prevents bees from getting into the sugar solution and drowning. Hive-top feeders can be left in position all year. A small bucketlike feeder is also available, which fit over the hole that is a feature of some inner covers, then a couple of supers are placed over to cover it. Another type, called a

ABOVE: An exposed hive showing the individual frames.

OPPOSITE: Checking the brood pattern.

Boardman or an entrance feeder, has an arrangement that enables an inverted jar of the syrup to be placed

BEES & BEE-KEEPING

Beehives in a field. Extra supers can be added as the hives fill up.

near the hive entrance, so that the bees can feed on it there. Although this system allows bees from other colonies to steal the syrup, it enables you to see how much food is being taken and to top it up without opening up the hive every time. Then there is a frame feeder, which replaces one of the normal types of frames suspended in the boxes.

OTHER EQUIPMENT

A few other items of equipment will be of help in your bee-keeping endeavours. First of these is a record book, because good record-keeping is important and may be of future value. For a start, it may remind you when you need to do certain tasks, or watch out for particular aspects of bee behaviour.

ABOVE: Checking the comb for larvae.

OPPOSITE: Still-life with smoker.

The things that didn't go to plan can also be noted, also the need for them to be revised. It is also useful to have details of the suppliers of your queen

KEEPING BEES

As you progress, you will probably find that a few items, such as a hammer, some pliers and a ball of string, will come in useful. It's worth keeping these together in a small tool bag or roll so that they are with you when tending the hive. Any other items proving their worth can be included as you go along.

BUYING YOUR BEES

In the same way that bee-keeping equipment can be obtained from specialist suppliers, so it will be when you acquire your actual bees. Rather like bee-keeping equipment, there are good bees and bad bees, so it pays to choose carefully. 'Bad' bees may be difficult temperamentally, making handling them a more irksome task than it should be, or they may come with brood diseases or may simply be inefficient gatherers, with the result that honey yields may suffer. Your first port of call might usefully be your local bee-keeping association, which will put you in touch with a bee-keeper in your area

and other bees to hand, as well as those of other suppliers, organizations, people you can contact in the event of an emergency or from which to seek general advice. The date each visit to the hive is made should also be recorded, together with weather conditions, temperature, and so on current at the time. At the same time, behaviour in the hive and the general temperament of the colony should also be noted down (for example, are the bees quiet or agitated?), while it may be useful to record the number of queen cells, the number of drones present and the general condition of the combs and of the hive itself. When making a visit, note down what tasks you undertook, including the amount of honey harvested.

Special equipment is required for collecting and storing honey (see pages 214 et seq.), but with regard to harvesting it, again it is valuable to keep a note of when this was done, the amount of honey collected, and any other relevant details, such as the price at which it was sold.

who has some good, surplus bees available, but first get some assurances regarding the quality of the stock and, if possible, some recommendations concerning the bee-keeper in question. The advantages of obtaining a colony in this way is that it will already be established and ready to go, and will be acclimatized to the local area. By the same token, however, it means that there is no 'starting-up period' whereby confidence can be gained as the colony grows – it's all action from the word go. Make sure these same bees come from hives which are at least 3 miles (4.8km) from the place where you intend to site your own hive, otherwise the bees may simply return to their original home when released from your hive.

The most popular course of action, however, is to obtain stock either from the supplier from which all or some of your equipment came, or from a recognized professional breeder of bees. The bees acquired from one of these sources are much more likely to be good stock, and a reputable breeder shouldn't object to having his bees inspected for disease by an expert before purchase. Of the different ways to buy bees, first, it is possible to buy a complete colony. This consists of ten or

LEFT: Never attempt to catch a natural swarm of bees yourself: leave it to the experts.

OPPOSITE ABOVE: You may purchase your bees in what is called a package containing honeybees and a queen.

OPPOSITE BELOW: Bees should by bought from a reputable supplier to ensure they are healthy and temperamentally sound.

so combs and contains a fertile queen, workers and, depending on the time of year, drones. The whole thing should come complete with stores of food and a brood (developing bees, including eggs, larvae and pupae). Obtaining such a colony in, say, May or June, should make it possible for the colony to produce surplus honey in the first year. The second method is to buy what is known as a nucleus, which is a small colony consisting of between four and six combs containing a fertile queen, some workers and perhaps some drones. It also includes some food and brood. A nucleus is a much smaller proposition than the complete colony described previously, and will be easier for the beginner to handle. Once installed within the hive the nucleus will, of course, develop into a complete

plenty of feeding or they will fail to survive the winter. The other issue regarding swarms is that, unless its origin, good health and good temper can be proven, it may turn out to be a bad group of bees.

RUNNING THE HIVE

Assuming the decision to keep bees has now been made, you will have read all that can reasonably be expected about the subject, you will have attended some bee-keeping sessions, and you have had all the necessary discussions with family and neighbours. You have probably even chosen the site and have invested in the

colony, and may even produce a small amount of honey in the first year.

A third route is to obtain what is called a package. This is a screened or meshed box simply containing honeybees and a queen. Once delivered from the breeder, you must transfer the bees into your own hive and feed them so that they get off to a quick start. The other option is to obtain a natural swarm of bees, which is a free-living colony without any combs. Early swarms tend to develop well since they have the whole flowering season ahead of them, but late swarms will require

necessary equipment, such as the hive, the tools, protective clothing and so on. Now, also assuming that it is spring or early summer, it is time to acquire your first batch of bees by one of the methods already described.

PUTTING YOUR BEES INTO THE HIVE

Let's assume, first of all, that you have bought a nucleus. The supplier may give instructions on how to transfer the nucleus and the frames to your hive

ABOVE: Healthy bees are more likely to produce plenty of honey. These are on a Langstroth frame.

OPPOSITE: The ultimate aim: delicious golden honey.

and how to provide any other immediate requirements such as food for the bees. Before taking delivery of the nucleus, check with the supplier so that you know what to do when it arrives. The essential procedures are as follows: when the nucleus arrives, place it on the hive stand, open the entrance and allow the bees to fly about for an hour or two. Next, take the nucleus off the hive stand and replace it with the hive. Remove the lid of the nucleus and give a puff or two of smoke from the smoker. Using your frame tool, prise apart the frames (you may need another puff or two of smoke to keep the bees out of the way as you do this). Lift each frame from the nucleus in turn and place it in the lowermost brood box, keeping the frames in the same sequence. Fill any empty frame spaces with new foundation frames. Put the inner cover on. A feeder with about 0.2–0.4 gallons (1–2 litres) of sugar syrup should be placed over the feed

OPPOSITE: These bees are well-established in their hives and are coming and going at will.

RIGHT: It is possible to buy an established bee colony which can be transported to your own hive.

hole before adding the outer cover. The queen excluder and the first honey super can be installed about a month later, assuming the colony is becoming established in full season.

Alternatively, you may have decided to purchase a package of bees. A typical package weighs about 3lbs (1.4kg) and consists of approximately 3,500 bees. The queen will probably be in a separate queen cage within the

package, and the worker bees will be clustered around the cage. Expect to see a few dead bees on the floor of the package box, but if the numbers seem excessive, contact the supplier. If the bees have exhausted their food supply, lightly spray a sugar solution (one-third sugar to two-thirds water in warm water) into the box using a mister. The bees must merely be moistened rather than saturated; don't leave them in the

box for any longer than is necessary once they have been delivered, but transfer them to the hive as soon as is practicable and providing the weather is not too cold. Late afternoon or early evening is a good time to do this.

Before putting the bees into the hive, make sure you have all your equipment to hand, that you are wearing your protective clothing, and that the smoker is lit. You may also need a pair of pliers. Remove the feeder from the package box, then remove the queen cage from the box. Remove any fixings from the feeder, then lift the box

and bang it on the ground so that any bees attached to the feeder fall off. Next, remove the queen cage and shake or blow off any bees attached to it. Make sure the queen is alive and looking healthy. At this stage, the queen cage can be inserted into the hive by pushing it between the frames in the centre of the lowermost hive box. Ensure that the mesh side faces downwards and that the workers can make contact with the queen. Within the queen cage is a section containing a sugar mixture that must usually be pierced with a pin or small nail.

Now place a second hive box on top of the first one, then remove some of the frames to make a space in which to dump the bees. Get the bees out of the package box by banging the box again, removing the cover and shaking or 'pouring' the bees downwards into the hive. Use the smoker to gently coax them down if necessary. You won't get every single bee out by this method, but when most of them are out, put the package box by the side of the hive so that the rest can come out in their own time, then carefully replace the missing frames from the hive box, trying not to trap any bees as you do so. Now replace the rest of the hive boxes and their frames and the inner and outer covers, attaching a sugar syrup feeder to the hive.

INSPECTING THE HIVE
Assuming sufficient food has been supplied when first inserting the bees, allow the colony to settle down for three days before inspecting the hive. On opening up the hive, first check the

LEFT: A queen excluder.

OPPOSITE: Always use paint that is non-toxic to bees on hives.

KEEPING BEES

activity of the bees surrounding the queen (don't use smoke before opening the hive because you want to observe the bees' natural reactions). If they seem generally calm, with only a few bees on the cage, it is probably safe to release the queen now. But if her cage is still completely surrounded by bees that are reluctant to be removed, close the hive up and wait for two more days.

When it seems safe to remove the queen, take off the plug covering the queen cage sugar mixture and make sure that the hole is clear by poking it with a pin or small nail. Replace the queen cage between the frames. In a few days, the workers will have eaten enough of the sugar mixture for the queen to be released. (Some bee-keepers simply release the queen

BELOW: After a few days, and when the bees have calmed down, the queen can be released.

OPPOSITE: Discovering eggs is a good sign that the colony is becoming established.

directly from her cage into the hive during the first inspection, but only if all seems well.) Refill the feeder and close up the hive. A couple of days

202

later, check that the workers have released the queen. If not, release her yourself by opening up her cage and letting her join the other hive members.

GETTING GOING

The queen should begin laying about a week after her release. The eggs are tiny white structures resembling miniature grains of sand, and they should be visible within some of the cells built by the worker bees. Discovering eggs is a good sign that all is well and that the colony is beginning to take off and get established. At this stage, keep the feeder topped up, but otherwise leave things alone for about ten days. If, however, no eggs are visible, despite your searches, first make sure that the queen is still present. If she is, it is possible that she wasn't mated before you got her, or that she is in some way unable to lay eggs. When this happens, the only course of action is to replace her with another queen.

As the colony develops and egg-laying proceeds, and as the workers continue to build their comb on the frames of the available boxes, you will need to add a new box of frames as the existing ones get built upon. It may also be necessary to swap around some of

the frames in the existing boxes so that they all get used. At this stage, keep using the feeder to supplement the bees' food supply. Make sure you always have sufficient supers ready for when the production of honey begins in earnest. The best advice is always to be aware of what is happening inside the hive with regard to brood, comb-production and honey storage, adjusting the number of frames you provide accordingly. External conditions should also be taken into account: for example, what is the

weather like? Is it conducive to bee activity? Are there plenty of local blooms available?

INSPECTING THE ESTABLISHED COLONY

There is often much to do when the hive is opened and, especially for the novice confronted by bees upset by the invasion of their home, it can be difficult to remember why you are inspecting the hive in the first place. Try to memorize the following before opening up:

• Does everything seem normal? Are the bees fairly active? Is there evidence of disease present? (Admittedly, such questions may only be answered adequately with experience.)

• Does the colony have enough food?

• Is all the space being filled? Is another super needed?

• Is the queen laying eggs? Is there older brood as well as eggs?

• Can you see new queen cells? Is the colony going to swarm?

The first task is to get the smoker lit and burning well. Have any equipment you may need to hand, such as the hive tool, together with the feeders and extra supers or brood frames you may wish to add. Don your protective clothing. Begin by introducing a small puff of smoke through the entrance door of the hive, then wait for a minute or so. Now

ABOVE: The presence of bee larvae is a good sign that all is well in the hive.

OPPOSITE: Inspecting the new colony.

you must remove the outer cover from the top of the hive. Lift an edge of the inner cover first; sometimes this will need a bit of encouragement, so use the end of your hive tool, puffing a little smoke into the gap you have made. Carefully lift off the cover completely.

Now you will see the inside of the topmost box. If the colony is still new, there may be little or no comb-building activity, but there will probably be some bees around. Lift a corner of the box and puff in some smoke. Now lift the box off completely and lay it carefully on the ground. Puff some smoke onto the top to prevent the bees from flying off. If the colony is going well, the next box you examine will probably show more signs of activity, particularly comb build-up in the centremost frames.

To examine a frame in detail, carefully lift it out of the box. It is best to begin by examining frames near to the outer edges of the box. Puff a small amount of smoke over the top of the box. Loosen the frame, if necessary, by gently levering or twisting the hive tool between two frames. Lift the frame out of the box, keeping it straight and taking care not to injure, trap or inadvertently kill any bees. If there is

no comb build-up on the frame, place it by the hive entrance for the time being to allow the bees to return to the hive. Follow the same procedure and lift out the next frame. If this frame shows signs of comb-building, check to see if there are any eggs. Hold the frame and its comb up over the hive (in case the queen is there and falls off) so that sunlight can shine on the cells, making the task easier. Then carefully replace the frame, taking care not to crush bees as you put it back and ensuring that it is positioned correctly. When replacing a box, it is better to slide it into position rather than dropping it onto a lower box. By doing this, you are encouraging the bees to move out of the way, avoiding crushing them. When you are ready, replace the inner cover, the feeder (if being used) and the outer cover. The whole process of examination should take no longer than 15 minutes or so once you have become proficient at the task.

OPPOSITE: Have your hive tool to hand when inspecting the hive...

RIGHT: ... also your smoker.

THE HIVE YEAR

The bees involve themselves in different activities according to the seasons, though these will naturally vary according to the hive's geographical location. For example, bee behaviour in temperate parts of the world won't be quite the same as that of bees living in the tropics, where warmer temperatures prevail. Below is a general account of the hive year.

In late winter, as the days begin to lengthen, the queen begins or increases her egg-laying activity. The workers consume the hive's stored supply of pollen and honey to produce food for the developing brood. In early spring the first flowers become available to help supplement the hive's stores of food, and the brood is beginning to increase rapidly. The expanding hive population may well trigger the start of swarming behaviour at this time. In temperate regions swarms usually occur from April to June.

By early summer, nectar and pollen are readily available in temperate regions, although they may start to decline in the tropics. Late summer sees another burst of nectar- and pollen-collecting activity in temperate regions. This is the period of major honey production.

When November arrives, most plant-flowering is over for the year and the colony starts to slow down. By winter the colony begins to cluster together to keep warm, the bees beating their wings to generate additional heat and feeding off their stored food reserves. On warmer winter days the bees may break their cluster to access the food in other parts of the hive, or go out on cleansing flights.

THE BEE-KEEPING YEAR

Whatever the time of year, there is always something for the bee-keeper to do. Here is a month-by-month summary of bee-keeping duties, although the timing of various tasks and events will, to an extent, be governed by local conditions. The record book should be kept up to date at all times.

ABOVE LEFT: In winter, the bees cluster together in the hive to keep warm.

OPPOSITE: In spring, the first flowers appear, ready to supplement the hive's dwindling food store.

KEEPING BEES

DECEMBER–FEBRUARY

- Check for hive damage caused by woodpeckers, squirrels or other animals and repair as needed.
- Check roof isn't leaking.
- Ensure hive entrance isn't blocked with debris or dead bees.
- Make sure the hive still slopes slightly forward to avoid rainwater accumulating on the top. Wedge the bottom of the hive at the back to ensure correct angle.
- Feed bees as necessary.
- Check condition of screening, hive tool, etc.
- Attend bee-keeping courses, meetings, and so on.

MARCH

- Increase feeding.
- Continue to check for hive damage and repair as needed.

APRIL

- Continue feeding with sugar syrup.
- Replace the floor with a clean one.
- Add a new super, if required, and fit the queen excluder.
- Apply varroa mite treatment (see also page 232).
- Be aware of early swarming activity.

MAY

- Regular inspections of the brood comb should start now. Replace old brood comb (about one-third of the brood comb will probably require replacement throughout the season).
- Check there is sufficient food in the brood chamber.
- Add any supers that may be required.
- Remove varroa treatment before honey begins to flow.

JUNE

- Check brood frames and replace any that are damaged.
- Continue to check for signs of swarming.
- Remove any frames that have wax-capped honey, and replace with new ones and/or additional supers.

JULY–AUGUST

- End of swarming.
- Remove queen excluder in August.
- Start to harvest honey in August.
- Reduce entrance with entrance block to prevent wasps from entering.
- Insert varroa strips for 42 days.

SEPTEMBER

- Begin feeding colony, using sugar solution with Nosema disease-killing agent such as Fumidil B.
- Remove varroa strips.
- Fit a mouse guard to the hive entrance.

OCTOBER–DECEMBER

- Ensure the hive is secure and cannot blow over in winter winds.
- Ensure entrance is cleared of dead bees (the mouse guard will restrict the entrance somewhat).

HONEY AND OTHER BEE PRODUCTS

Bees may be kept simply for the pleasure of caring for them, becoming involved in their world, and to help to increase declining honeybee populations; but chances are that the main reason is to have a ready supply of honey – and perhaps beeswax. In the same way that there are well-established rules for tending bees, it is important to follow the correct procedures when harvesting honey. Remember that a foodstuff is being handled and prepared, some or all of which may be sold on to the public. You must therefore ensure that everything is kept scrupulously clean and that regulations applying to the sale of honey in your area are strictly adhered to.

HARVESTING HONEY

Numerous books and websites are devoted to the subject of harvesting honey, extracting and putting it into

jars, but below is a general account of the process, which describes the main stages and the options available. Honey extraction is a somewhat messy business requiring a few pieces of specialized equipment, some method of temperature regulation, and a little know-how. Should you decide this aspect of bee-keeping is not for you, other bee-keepers may perform the task for you.

You will know when it is time to begin collecting the honey because the supers have their honeycombs closed off with wax coverings. Get your smoker lit, put on your protective clothing and have your hive tool to hand. The first job is to get all the bees out of the super, or supers, from which

Well-sited and -maintained beehives.

KEEPING BEES

the honey is to be obtained. To facilitate this, a bee escape is used. This is a board that is placed under the super you wish to remove. It contains an exit that allows the bees to move down into the brood area, but prevents them from returning to the honey super. Give the bees 24 hours to vacate the super. Some bee-keepers use a fume board to encourage them to leave; this is a special board with a cloth impregnated with a safe chemical, such as benzaldehyde or butric anhydride, the first of which smells like bitter almonds, and which can be obtained

BELOW and OPPOSITE: Honey is stored in the honeycomb cells.

from bee-keeping suppliers. When a fume board is used, it normally takes only a quarter of an hour or so for the bees to leave; the chemical is not poisonous to them, they merely find it offensive. Another method is to use a mechanical blower to forcibly remove the bees from the super, having first removed it from the hive. The super should be placed on top of the hive with the bottom of it facing the back of the hive before a blast is directed through the frames. In any case, a light smoking is useful to begin the evacuation process.

EXTRACTING AND BOTTLING HONEY

Once the bees have vacated the box, the frames containing honey can be removed. Honey extraction should be carried out in a clean room using clean

BELOW: The honey is ready for harvesting when the honeycomb has been capped off with wax.

OPPOSITE LEFT: A disassembled bee escape.

OPPOSITE RIGHT: The bees can be encouraged to leave the super by using a blower. The novice is advised to wear gloves.

materials; the honey will flow better if the temperature is warm. In many ways

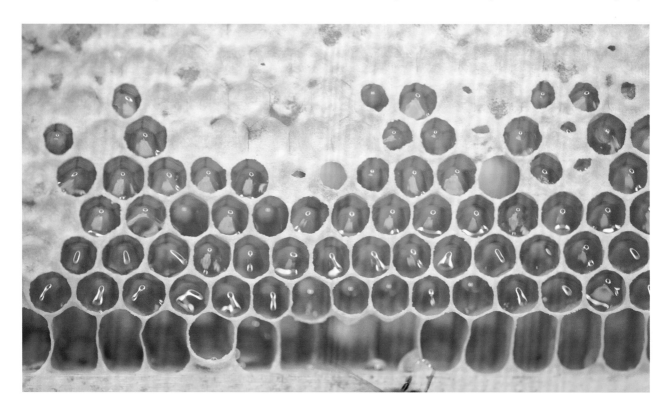

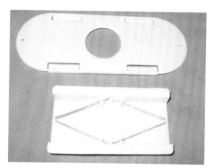

a kitchen is the ideal place, being warm and with access to power and running water. This may not accord with everyone's idea of what a kitchen should be used for, however, and so a utility room or even a garage may have to double up as an extraction room, provided it is not adjacent to a lavatory. Hands must be thoroughly clean, with waterproof dressings covering cuts, and clean protective clothing must be worn. Clean up as you go along.

All equipment must be of food-grade plastic or stainless steel. First get a bucket and place some cheese cloth inside it, allowing some of the material to drape over the edges. Now place a piece of wood approximately 3 x 2-in (7.5 x 5-cm) thick across the top of the bucket. A couple of notches cut out to make it fit the rim of the bucket without slipping around, and another on the opposite side to hold the frame securely, will greatly assist the task that follows. Now, resting one end of the frame on the piece of wood, cut or carve off the wax cappings that seal the comb, leaning the frame slightly at an angle so that the cappings fall off without sticking back onto the frame lower down. A proprietory uncapping knife (some are heated to make the task easier) or a sharp, strong carving knife are suitable for the purpose, using the two edges of the frame as cutting guides. You will see the capping fall into the bucket. The honey that flows out with the cappings can be gathered once it has drained through the cloth. Once you have completed one side, turn the frame over and do the same with the other side. Once the process is understood, more efficient ways of collecting and straining the honey may be devised – perhaps a plastic box with a fine filter can be used, set above a second box with a tap built into the side near the bottom to allow the collected honey to flow out.

The honeycomb should now be placed in an extractor, of which there are several types, chief among them being the radial and the tangential, named according to the way the frames are held in the unit. Both, however,

extract the honey by using centrifugal force – rather like the action of a spin-drier removing water. The best extractors are made from food-grade polythene or stainless steel; other materials are not suitable for honey that is intended for public consumption. Some extractors are driven by electric motors and others are operated by turning a handle. It is also possible to hire a good extractor from bee-keeping associations rather than buying one. Always follow the manufacturer's instructions, paying attention to the way the extractor is loaded and cleaned after use.

The honey should then be finely strained directly from the extractor, a task made easier if it is warm, bearing in mind that once the honey starts to granulate it will not pass through a sieve unless it is warmed. The strained honey is collected in a bucket prior to bottling. The best way to remove honey from the bucket is by means of a tap called a honey gate, and it is an easy

OPPOSITE: Remove the wax cappings from the frame using a specialist knife or a carving knife.

RIGHT: The honey extractor uses centrifugal force to remove the honey from the combs.

task to fit one yourself, nylon honey gates, with the necessary washers and fixings, being readily available. Cut a hole near to the bottom of a plastic bucket and fit on the honey gate. Again, you will need to warm the honey so that it flows easily out of the bucket and into jars. If you are competent at DIY, a simple way to do this is to build a warming box comprising a couple of light bulbs (each about 40W). Place the buckets on the warming box until the honey is of the desired consistency but without overheating it.

Jars must be perfectly clean and dry and preferably sterilized by filling each clean jar with water and microwaving it on full power for five minutes, depending on the size (make sure there are no metal parts on the jar). Empty, and allow to cool. Low, wide-necked jars are preferable so that the honey can be more easily removed. Once the honey has been poured into the jars they should be carefully sealed to prevent moisture absorption, which may cause fermentation, then labelled and dated. Attractive labels may be

LEFT & OPPOSITE: The honey is drained from the extractor via a strainer.

purchased for a more professional look. Unless you produce honey on an industrial scale, using bees that have been restricted to specific flowers (such as heather), it is unlikely you will be able to identify the type of flower from which it came, for during the course of a season the bees will have collected nectar and pollen from all manner of flowering plants. This will be very evident once the honey extraction process is begun; the colour of the honey may vary considerably, reflecting the shades and colours of the flower products from which it originated.

Having such a high sugar content, and being a natural antibacterial and antifungal, honey is considered a low-risk food. The above method is primarily intended for use at home when preparing honey for your own personal use. Some beekeepers sell honey only occasionally in small and variable amounts according to the size of the harvest and the honey that is surplus to their requirements each year. Others, with a greater number of hives, may regard their bee-keeping as a sideline to earn extra money and may be supplying various retail outlets on a regular basis. Remember that the rules are rather more stringent for bee-

keepers supplying honey for sale to the public. In the United States, the Federal Food and Drug Administration or US Department of Agriculture should first be approached, while in England HM Government's requirements covering production, labelling and lot numbering should be consulted (see internet). (Other parts of Britain have their own statutory requirements.) Statutory regulations change from time to time, and your local Environmental Health Officer should also be approached for advice. Advisory leaflet No. 103: *So You Wish To Sell Honey*, also attempts to set out and clarify these requirements, and may be obtained from: BBKA The Bee Centre, National Agricultural Centre, Stoneleigh, Warwickshire, CV8.

USES OF HONEY

This may seem an obvious and unnecessary section to include. After all, honey is for eating isn't it? Well, yes, of course it is, but honey has more to

OPPOSITE: The honey can be poured into sterilized jars once it has been strained.

RIGHT: Honey varies in colour depending on the flowers or plants that contributed to its makeup.

offer than simply being a form of food. Mention was made earlier of the value of honey as a trading commodity, as a natural preservative and as the basis for the alcoholic drink known as mead, but honey also plays a large role in medicine. For 2,700 years, and maybe even longer, humans have used honey to treat all manner of ailments, applying it to wounds, for example, to combat infection and speed the healing process, although it is only fairly recently that the antiseptic and antibacterial properties of honey have been fully explored. Honey is now approved as an agent to help combat the dangerous MRSA bacteria, and it is also used in the treatment of diabetic ulcers. The antioxidants present in honey have also been attributed to alleviation of conditions such as colitis. Patients, after their tonsils are removed, are prescribed honey, and indeed honey has been used for centuries to relieve sore throats and coughs, either taken in liquid form, and often mixed with lemon and water, or in the form of honey lozenges.

Cookery books are packed with recipes using honey; indeed, many such books are devoted solely to the art of cooking with this ingredient. Honey

adds sweetness, body, a unique flavour and a delicious glaze to many dishes, and goes especially well with meats such as pork (and ham), chicken and duck, as well as with fish such as salmon. Food can be coated with honey or it can be one of the ingredients of a marinade or sauce for meat or vegetables. It is often used to top waffles, fruit salads, breakfast cereals and yoghurts, or as a spread on bread or toast.

Apart from its use in making mead – of which there are numerous varieties – honey can be used in several other types of drinks, including mulled wines. A particularly effective remedy to ease sore throats and relieve the affects of colds and flu can be achieved by sipping a mixture of two teaspoonfuls of honey with the juice of half a lemon from a cup topped up with boiling water. Adults may find this drink even more beneficial if a small tot of whisky is also added, especially at bedtime!

OPPOSITE: Honey has long been known for its beneficial qualities.

RIGHT: A honey and lemon drink works wonders as far as colds and flu are concerned.

OTHER BEE PRODUCTS

Another useful commodity extracted from bees is the substance known as royal jelly. This, as we have seen, is a secretion produced by the hypopharyngeal glands of worker bees and fed to developing larvae. Royal jelly is produced commercially by stimulating the colony to produce queen bees, from which the royal jelly is collected when the larvae are only a few days old. It is practical only to collect the royal jelly from developing queens, for although other larvae are also fed the substance for a few days, only queens receive a store of it that can be collected. During the processing of royal jelly, honey and beeswax are also added to help in its preservation.

Royal jelly is sold as a dietary supplement, and it is claimed to have various health benefits, due mainly to its high vitamin content, especially the B complex. Royal jelly may also have some value in boosting the immune system, in the stimulation of stem cells in the brain, in lowering cholesterol, and as an antibiotic and anti-inflammatory – properties that are

OPPOSITE & RIGHT: Honey and its byproducts have proven their worth over the centuries.

unlikely to be fulfilled if the product is ingested, when they are neutralized. Royal jelly is also a component of some beauty products.

Beeswax is another natural product, secreted from special glands on the abdomens of worker bees. The wax is used for building the comb cells in which the young are raised and the pollen and honey stored. Beeswax is also variously used around hives to fill in gaps. During the honey extraction process (see page 216), the wax cappings are cut from the comb. The colour of the wax varies according to the types of flowers on which the bees were feeding, but it is generally yellow, although it can vary from nearly white to brown. The wax is sieved off and gathered during the honey extraction process to be purified before being put to a variety of uses. As well as the many historical uses of beeswax, which included the making of candles, seals and sculptures, clarified beeswax is still used for candle-making today, as a lubricant in the woodworking and

OPPOSITE & RIGHT: Beeswax has been used in candle-making for thousands of years, the candles, when lit, producing a pleasing natural fragrance.

cabinet-building trade for the smooth operation of drawers and windows, and in wood and shoe polishes when dissolved in turpentine.

Beeswax is occasionally used as a coating for cheeses and in the cosmetics industry (for example, as hair pomade), and is used in medicine to make dentistry casts and barrier creams, the cosmetics and pharmaceutical industries accounting for more than 50 per cent of the total consumption. It is quite possible for bee-keepers to collect and refine beeswax for use in some of the aforementioned, and there are many books and websites that explain in detail how this may be achieved.

Honey is one of Mother Nature's finest inventions, for besides its historical use as a natural sweetener it is also seen as a panacea for countless ills, containing as it does many essential and life-giving nutrients.

There are up to 200 different substances in honey, including fructose, glucose, vitamins and minerals, proteins, amino acids and enzymes. Honey is the product of one of the world's most efficient factories – the beehive – and bees may travel as much as 55,000 miles, and visit more than 2 million flowers, to gather sufficient nectar to make a mere pound of honey.

BEES & BEE-KEEPING

Honey was the first sweetener, predating sugar by hundreds of years. It follows, therefore, that it has many of the same culinary uses as sugar, while at the same time having unique properties of its own. It is as sweet as sugar but infinitely tastier, adding that something extra to barbecued and marinated foods, cakes, cookies and fruit desserts, and especially oriental cuisine as the sweet element married with sour, bitter, salty, hot, pungent or spicy. Should you need the runny variety for a recipe, and find you only have set honey in the cupboard, place the required amount in a suitable dish and microwave for 30 seconds or so when it will become runny.

CHAPTER NINE

ENEMIES OF BEES

In the wild, bees must contend with all manner of predatory wildlife intent either on eating them, parasitizing them, or raiding their nests to steal their honey. Hive bees in their little wooden fortresses are more immune to some attacks, but they still fall prey to a variety of specialist bee-eating animals when out on their forays. Furthermore, they are under threat of attack from a variety of diseases and pests that target bees, as well as from interlopers that have decided that a beehive is an excellent place in which to make a home. Some ailments affecting bees are merely troublesome, but others may be lethal if left untreated. Many of these afflictions, however, can be reduced or eradicated as part of a regular programme of good management. Most pests and diseases can be eliminated, given appropriate treatment regimes, but something occasionally takes a hold that sadly can only be eliminated by the destruction of the hive and its inhabitants. The afflictions of honeybees include pests and parasites, bacterial, fungal and viral diseases, as well as some directly related to other factors, such as sudden changes in external weather conditions.

VARROA

Varroa is the name given to species of parasitic mites, such as *Varroa destructor* and *Varroa jacobsoni*, which feed off the body fluids of bees during several stages of their life cycles. The

OPPOSITE: Varroa mites on bee larvae.

RIGHT: The small hive beetle (*Aethina tumida*).

adult mites are clearly visible as tiny brown or red spots on the adult bee's thorax, or on the larva, when they are especially easy to see. Varroa carries a virus that can damage bees, and bees infested thus will also show other symptoms, such as deformed wings and reduced vigour. Varroa has spread slowly through most continents of the world, having reached the United States in the 1980s, and been detected in Britain in the early 1990s. Left unchecked, varroa can lead to the destruction of colonies, although there is evidence that some colonies are developing a resistance to the pest.

Several treatments are available. Chemical controls, when used as directed, are able to kill large numbers of mites without unduly disrupting a honeybee's activity, but bear in mind that anything that can kill mites is clearly toxic, and in time a build-up of the active ingredients may get into the beeswax and begin to affect the bees as well. Chemical controls are usually applied on strips that are placed in the hive and hung between the frames, but

which must be removed later on according to the instructions. Another treatment is sucrose octanoate, often sold as 'sucrocide'. The substance, a sugar-and-soap solution, is sprayed directly onto the bees, the treatment needing to be carried out several times. It works by clogging up the mites' breathing apparatus but not the bees', and is probably the safest treatment.

TRACHEAL MITES

Acarapsis woodi is the parasitic acarine mite responsible for infesting the airways (tracheae) of honeybees. Unlike the varroa mite, it is microscopic in size and is more or less universal in its occurrence; the chances are that bees that are bought will already have these mites unless they are specifically marketed as being tracheal mite-resistant. Mature female acarine mites leave the tracheae and wait on the body of the bee until they can transfer to another victim. Then they move into its airways and begin laying eggs. Left unchecked, mite concentrations can build up in the hive, reducing honey production, and most bees will die in the winter.

Treatment is by the introduction into the hive of so-called grease

patties. These are made using one part vegetable shortening to three or four parts powdered sugar and are placed on the top bars of the hive. The bees then eat the mixture and in so doing pick up traces of the shortening, disrupting the mites' life cycle. Make sure to keep a grease patty in the hive from autumn through to the following spring, replacing it when it has been consumed.

NOSEMA

This spore-forming parasite invades the intestinal tracts of adult bees. It usually becomes apparent when bees cannot leave the hive to eliminate waste (during, for example, a spell of prolonged winter weather). Nosema

ENEMIES OF BEES

WAX MOTH

The wax moth or greater wax moth, *Galleria mellonella*, can be very disruptive in a hive, but fortunately its removal is fairly straightforward. A female wax moth must first get into the hive by slipping past the bees guarding the entrance. Once inside, she lays eggs in one of the brood boxes. The emerging wax moth larvae, or waxworms, then proceed to eat beeswax, pollen, honey and bee larvae and pupae. Sometimes, the hive bees eject the interlopers, but at other times they seem able to secure a foothold. As the larvae burrow through the combs they leave a trail of webbing that snags the bees. Then they pupate, forming tough cocoons that are difficult to remove. When the adult moths emerge, they leave the hive, mate outside, then look for more hives to infest. Wax moths can also infest supers that are stored inside or left outside in mild climates because they are not in use at the time. Fortunately, freezing conditions kill wax moth eggs and larvae. Storing supers in cold, fresh air, but protected from the rain, is a good preventative measure against infestation by wax moths.

(*Nosema apis*) can be treated by increasing the ventilation in the hive or by using antibiotics.

SMALL HIVE BEETLE

Aethina tumida is a tiny dark beetle inhabiting hives. It originated in Africa, but has only been a problem in the western hemisphere since the late 1990s, when a specimen was identified in Florida. Adult female beetles enter the hive to lay their eggs, and when hatched, the beetle larvae work their way through the comb devouring bee larvae, pollen and honey. Mature beetle larvae then leave the hive, pupate in soil nearby and emerge as adults to infest another colony or to re-infest the original one. High levels of infestation may well drive out the resident bees.

Treatment initially involves removing any infested supers from the hive, but other measures include pesticides that can only be accessed by the beetles in screened bottom boards and bottom-board traps. Some bee-keepers use a chalky powder – diatomaceous earth – around the hive as a method of interfering with the beetle's life cycle. This is a relatively new pest, and more effective control measures will undoubtedly be developed in time.

Normally, a non-stressed, healthy colony takes care of its own wax moth problems, removing the larvae and clearing up the mess they leave behind. Some chemical treatments are also available.

AMERICAN FOULBROOD (AFB)

This is caused by the bacterium *Paenibacillus larvae*, and is the most widespread and destructive of honeybee brood ailments. Bee larvae, in their first three or so days of life, become infected by eating spores that are present in their food; older larvae appear immune to the disease. The infected larva, sealed within its cell by the worker bees in the normal manner, then dies, although its body may carry up to 100 million spores. When cleaning out the infected cells, the bees then distribute the deadly spores throughout the colony, and contaminated brood food then infects other larvae. A thus weakened colony may be invaded by robber bees, which then take the disease to other colonies. Signs of AFB include a spotted pattern in the brood comb and discoloured, sunken or punctured wax cappings. Scales, the dried remains of dead larvae in the cells, may also be visible. Bee-keepers may inadvertently transmit the disease when they move equipment from infected hives to healthy ones. Fortunately, the disease is rare and control measures can be implemented quickly should it occur.

Chemical treatment is available, and many bee-keepers routinely apply a preventative drug whether AFB is suspected or not. The drugs, including terramycin antibiotic, are applied to the broodnest with powdered sugar. One of the problems with this course of action, however, is that when an outbreak of AFB occurs, the treatment needs to be applied more or less in perpetuity. The route taken by many bee-keepers, although drastic, is to destroy by burning everything associated with the outbreak – the bees, the hive, and all equipment that has come into contact with the hive and the bees.

Alternatively, it is possible to burn the frames and comb but flame-scorch the interior of the hive instead before treating with a disinfectant. Before any such measures are taken, however, it is worth seeking professional advice as to the best way forward. Meanwhile, it is worth stressing the importance of cleanliness when using equipment. It is also good practice never to use second-hand equipment, such as hive tools, since there is no way of knowing if they are infected with AFB spores.

EUROPEAN FOULBROOD (EFB)

European foulbrood is caused by the bacterium *Melissococcus plutonius*. It is far less deadly than American foulbrood, and is described as a 'stress' disease – in other words, a disease that becomes dangerous when the bee colony is already weakened or under stress for other reasons. It attacks very young larvae, passed to them by nurse worker bees that inadvertently spread the disease throughout the colony when tending the young. The larva eats food containing the bacteria, which take up residence in the host's gut, competing with it for food and subsequently killing the bee larva. Scales (the remains of dead larvae) may be visible in some of the brood cells, and the brood comb may develop a random pattern containing holes.

European foulbrood is treated using the same method as for American foulbrood. Generally, if a bee-keeper is treating the colony for AFB, then EFB is not a problem. Avoiding stress in the colony is also a factor in helping to combat the disease: keep food levels up, maintain a strong colony, and replace old frames when necessary.

CHALKBROOD

A fungal disease caused by *Ascophaera apis*. When a bee larva takes in the spores, the fungus grows in the larva's gut and competes with it for food, eventually starving it. In time, the fungus envelops the whole larva and also fills the larval cell. White, mummified larva can be seen in cells in the brood comb and also outside the hive, where they have been ejected by members of the colony.

Replacing the queen and removing the infested combs is the first course of action. It may also be worth replacing entire frames if infected, and replacing them every three years or so in any case. Wet spring weather may encourage chalkbrood to develop, so good hive ventilation is important. Some authorities recommend disinfecting with 80 per cent acetic acid.

STONEBROOD

This is another, less common, fungal infection known to affect the honeybee. Stonebrood is caused by fungi such as *Aspergillus flavus* and *Aspergillus fumigatus*. It infects larvae in a similar way to chalkbrood, but turns the infected larvae black when they die. In a healthy colony, workers are usually able to clean out the infected cells themselves.

CHILLED BROOD

Not actually a disease but a condition that can be brought on by mismanagement of the hive. It can sometimes also occur as a result of inadvertent contact with insecticides or a sudden drop in the outside temperature during the rapid spring build-up. When this happens, the colony clusters in the middle of the hive to keep warm, and uncapped brood on the edge of the brood comb, which are normally kept warm by the workers, will die from exposure. It is difficult to prevent this, but opening the hive during the day in warm weather will help raise the internal hive temperature.

VIRUSES

Several viruses are known to cause diseases in bees. It is worth listing the best-known of them here, although more work needs to be done before they are fully understood. It appears that many of these viruses are always present in colonies but only manifest themselves when the colony is stressed or other ailments have taken hold. They include the acute paralysis, chronic paralysis, sacbrood, Kashmir bee, black queen cell, cloudy wing and deformed wing viruses.

DECLINING BEE POPULATIONS

Much has been reported in recent years concerning the decline in the numbers of bees in many parts of the world. This is a very worrying trend, since bees and other pollinators play such a vital role in food production. In the United Kingdom, for example, bees pollinate many crops, such as fruit and vegetables, and are worth many, many millions of pounds each year to the economy. After the winter of 2008, it is estimated that one in three of the nation's hives failed to survive, the highest losses being in the north. This figure compares with previous averages of around only 7 to 10 per cent losses. In many parts of the United States, feral honeybee populations have dropped by about 90 per cent over the past 50 years, and managed honeybee populations have fallen by about two-thirds.

Possible explanations for the decline are numerous. They include:

• Increases and overuse of pesticides, domestically as well as commercially, particularly when plants are in bloom and therefore frequently visited by bees.

• The rapid transfer of diseases and parasites around the world. The global economy means that goods and commodities travel quickly to many foreign markets, sometimes bringing pathogens with them. American foulbrood, varroa mites and African hive beetles are among several diseases spread in this manner.

• Changes in agricultural practices, such as hedgerow removal, destruction of flower meadows, monoculture (including replacing deciduous woods with pinewoods), removal of plants from waysides, and general loss of habitat, all resulting in fewer feeding opportunities for bees and other pollinators.

• Air pollution. The fumes generated by motor vehicles and industrial plants have the ability to mask the fragrance of flowers, causing bees to travel further in their search for food.

You are helping the environment and improving the decline in bee populations by keeping bees in your backyard.

GLOSSARY

Like many somewhat specialized activities, bee-keeping has its own 'language'. Most of the special words or phrases relating to bees and bee-keeping are described in the text the first time they are mentioned, but below is a glossary of the key words and their meanings.

Abdomen The major body region amounting to a third of an insect such as a bee. Within the abdomen of a bee are found the honey sac, the intestine and some other parts of the digestive system. The abdomen also contains the reproductive organs, much of the respiratory system, the sting, and glands such as the wax glands and the Nasonov gland.

African honeybee A variety of honeybee created through the crossing of an African and a European subspecies of honeybee. It is regarded as extremely aggressive in defence of its hive.

Alarm pheromone Released by a worker bee when she stings, for example, in defence of the colony.

Anther The part of the flower that produces the male reproductive cells, or pollen.

Apiarist Another name for a bee-keeper.

Apiary The place where honeybee colonies and hives are kept.

Apiculture The keeping and rearing of bees, especially for their honey.

Apis mellifera The genus and species of the European or Western honeybee.

Bacterium One of a large group of single-celled parasitic or saprophytic micro-organisms.

Bee space The areas of a hive where bees are free to live and work and will not block with wax, especially the spaces between combs. The bee space is between $1/4$ and $3/8$in (5–8mm).

Beeswax A special compound secreted from eight glands on the worker bee's abdomen. It is used for creating six-sided cells to form the comb. Bees also use wax to seal up gaps in the hive or nest.

Brood Describes the immature or non-adult stages of bees' development, i.e., eggs, larvae and pupae.

Brood chamber The part of the nest or hive in which the brood is reared. Also called the brood comb.

Cape bee A dark-coloured species from coastal regions of South Africa.

Capped brood Those pupae whose cells have been sealed with a wax covering during their non-feeding period.

Carniolan A variety of honeybee from eastern Europe.

Caucasian A variety of honeybee from Eurasia.

Cell One of the six-sided waxy compartments of a honeycomb built by worker bees and used for rearing young or storing food.

Cluster When a group of bees huddle together for warmth.

Colony The adult bees and their brood that live together in a nest or hive.

Comb A sheet of six-sided cells made from beeswax in which honey and pollen are stored or in which the brood is reared.

Comb honey Honey that is sold while still on the comb.

Compound eye One of the paired sight organs on an insect's head, and which is composed of many small units called ommatida.

Crop *See* Honey sac.

Dance One of the various dancelike movements made by worker bees to signify to other hive members the location and quality of a food source. *See also* Waggle dance.

Drawn comb A comb with cells built out by bees from a foundation sheet.

Drone A male honeybee.

Drone comb A comb in which the queen lays eggs that will develop into drones.

Egg The first stage in the life cycle of a honeybee.

Entrance reducer A device, usually made of wood or metal, used to restrict the hive entrance so that the hive is easier to defend against robber bees or to reduce its exposure to the elements.

Extracted honey Liquid honey taken from the comb by means of an extractor.

Extractor A device used to remove honey from the comb.

Fanning The method by which worker bees send the Nasonov pheromone out from the hive as a beacon to guide incoming bees.

Feeder Any one of several devices designed to feed sugar syrup to honeybees.

Fertile queen A queen with the ability to lay fertile eggs.

Fertilization The union of a male and a female sex cell.

Foraging The act of looking for, and collecting, pollen, nectar, water and propolis by bees outside of the hive or nest.

Foulbrood An infectious disease affecting honeybees that is caused by a bacterium.

Foundation sheet A proprietary sheet of beeswax or plastic embossed with cell bases on which the worker bees build their waxy cells. Also known as a comb foundation.

Frame A four-sided frame or structure designed to hold a foundation sheet and drawn comb.

GLOSSARY

Hamuli An arrangement of tiny hooks found on the leading edges of a bee's hindwings that attach to the trailing edges of the forewings to hold the wings together in flight.

Head The first of the major body parts of an insect such as a honeybee. It contains the eyes, antennae, mouthparts and various other structures such as glands.

Hive An artificial (man-made) home for bees. Some of the most popular hives include the Langstroth, the Commercial and the National.

Hive tool Used for opening hives, separating frames, and removing wax and propolis.

Honey A sweet substance made by bees from the nectar of flowers and used by them as a source of food. It is composed mainly of the sugars glucose and fructose, but also contains small amounts of sucrose, water, enzymes and minerals. Honey is also harvested by bee-keepers.

Honey flow The period during which nectar is available and bees are actively making honey.

Honeyguide A structure on a flower that helps lead insects to nectar. Also a type of bird that specializes in feeding on the larvae in bees' nests.

Honey sac The part of the digestive system of a honeybee that is used for transporting nectar, pollen and water back to the hive.

Honey surplus Honey stored in the hive that is not required by the bees and can be taken by the bee-keeper.

Honeycomb The part of the nest or hive that is used by the bees for storing honey.

Hymenoptera The order of insects into which the bees, wasps and ants are grouped.

Hypopharyngeal glands Found in the head of worker bees and which produce the substance known as royal jelly. Also known as brood food glands.

Inner cover Fits over the uppermost box in a hive.

Italian bee A popular and widespread variety of honeybee.

Larva (plural: larvae) The legless, grublike feeding stage in the development of an insect's life cycle.

Mandibular gland Found in the head of the queen bee and used to secrete a pheromone known as queen substance; the presence of this maintains the correct social organization within the colony.

Marked queen A queen bee sold by a breeder that has been marked with a spot of paint on the thorax. The paint makes it easier to find the queen in the hive and also ensures she is the original queen and not a different one.

Mating flight Undertaken by a virgin queen during which she mates with one or more drones.

Metamorphosis The changing stages in the life cycle. In a honeybee, this consists of egg, larva, pupa and adult.

Micro-organism An organism too small to be seen without a microscope.

Nectar A sweet liquid secreted by many flowering plants in order to attract insects.

Nucleus (plural: nuclei) A small frame hive (usually consisting of between two and five frames) acquired for starting a new colony.

Nurse bee A bee of about three to ten days old that feeds and cares for the developing brood.

Ocellus (plural: ocelli) Non-image-forming eyes situated on the heads of many insects that detect light and assist in orientation. A honeybee has three ocelli.

Outer cover The topmost, weatherproof cover on a hive.

Ovipositor The often tubelike egg-laying apparatus of many animals, for example, the honeybee.

Package (of bees) A screened or mesh-covered box containing some honeybees and a queen, acquired with the intention of starting a colony.

Pheromone A secreted chemical used to convey a message from one animal to another. Honeybees use several different pheromones as a way of communicating messages throughout the colony.

Pollen Male reproductive cells produced by flowers and used by bees as food.

Pollen sac A specialized region on the rear pair of legs of some bees, such as honeybees, that is used to transport pollen collected from flowers back to the hive.

Pollination The act of conveying pollen from the anther to the stigma, ovule, flower or plant to allow fertilization. Insects such as honeybees play a vital role in this activity.

Proboscis The tubelike, sucking mouthparts of an insect such as a honeybee.

Propolis A substance used by bees to strengthen the comb, seal up cracks, and reduce entrance holes in hives. It is made from the gathered sap and resins of certain plants mixed with bee enzymes.

Pupa (plural: pupae) The third stage in an insect's life cycle, during which the insect turns from a larva into an adult.

Queen A special type of female bee (and larger than a worker bee) that is capable of reproduction.

Queen cage A small, boxlike cage used for transporting and introducing a queen into a colony.

Queen cell An elongated cell in which a queen is reared. Queen cells hang down vertically from the bottom of the comb, as opposed to worker and drone cells, which are smaller and arranged horizontally.

Queen excluder A metal or plastic grid that fits into the hive and allows the passage of workers through it while restricting queens or drones from certain parts of the hive.

Queen substance A pheromone produced and secreted by a queen bee. The correct level of the pheromone maintains stability throughout the hive.

Rabbet One of the ledges on the inner, upper edges of a hive box from which the frames are hung.

Requeen To replace the colony's queen with a new queen.

GLOSSARY

Robbing The act whereby bees steal honey from another hive.

Royal jelly A very nutritious food produced by the glandular secretions of worker bees. It is fed to all larvae for a few days, then fed thereafter only to larvae destined to be queen bees.

Scout A bee whose job it is to look for food, water or propolis. Scout bees also prospect the area for a new home for the colony when it becomes necessary.

Smoker An apparatus used to make smoke which calms the bees when bee-keepers are working with a colony.

Spermotheca (plural: spermothecae) An internal organ found in the queen that stores sperm from a drone. The queen releases sperm from the spermotheca to fertilize an egg just before it is laid.

Stamen One of the usually filamentous structures that support the anthers on a flower.

Stigma The top part of the female reproductive system on a flower, on which the pollen lands prior to fertilization.

Sting The modified ovipositor of an insect, such as a honeybee, that is used by workers to defend the nest or hive, and by the queen to kill rival queens.

Style The usually elongated part of the female reproductive system on a flower.

Super A hive box or frame placed above the brood chamber where worker bees store surplus honey.

Supersedure The replacement of a queen by a daughter in the same hive.

Swarm A group of bees, consisting of approximately half of the workers, a few drones and the queen, that depart the original nest or hive to form a new colony.

Thorax The middle part of an insect's body on which are found the legs, the wings (in winged insects) and most of the body muscles used for walking and flight.

Tracheal mite A parasite known as *Acarapis woodi* that infects the trachea of a honeybee.

Uncapping knife A special tool used for removing the cappings from sealed honeycombs.

Varroa mite A parasitic mite, *Varroa destructor*, that infects honeybee pupae and adults.

Veil The meshlike curtain attached to the headgear of a bee-keeper's suit to keep bees away from the face.

Venom The chemical injected into the skin via the jaws or the sting of certain animals. In honeybees, the venom is injected via the sting.

Virgin queen An unmated queen bee.

Virus One of many microscopic, often disease-causing organisms that can only reproduce and carry out their life cycle inside another cell.

Waggle dance A series of figure-of-eight movements performed by bees to inform other bees of the location and quality of food outside the hive.

Wax moth The larva *Galleria mellonella* that causes damage to brood combs.

Worker bee A female with undeveloped reproductive organs. She performs most of the duties in the colony apart from laying fertile eggs.

INDEX

BEES & BEE-KEEPING

INDEX

ACKNOWLEDGEMENTS

Photographic Acknowledgements
Front Cover: iStockphoto LP © Marco Harzing.
Back Cover: Flickr/Creative Commons www.creativecommons.org Ian Sutton.
Spine: iStockphoto LP © Kelly Kline.

Flickr/Creative Commons www.creativecommons.org:- Alan Vernon: page 129. Annia316: page 52 below right. Aunt Jojo: page 208. Aussiegall: pages 52 left, 79 below, 82 above. Beatrice Murch: page 91. Becca G: page 10. Ben Feeder: page 192. Ben Klucek: page 58 top left. Ben Ostrowsky: page 47 above. Benson Kua: page 20 left. Bill Tyre: page 29. Bixentro: page 158. Bob Gutowski: page 59 top right. Bob Tie GuyII: page 116. Brian Giesen: page 57 top right. Brigitte Wohack: pages 103, 107. 00hCaffiene: page 46 right. Carly & Art: pages 12 right, 130 left. Castor_girl: page 55 below. Cello8: page 55 top left. Chris in Wales: page 136. Clare Bell: page 58 top right. Colin Howley: pages 62-63, 209. Dain Neilsen: page 55 top right. Daniel Feliciano: pages 135, 249. DDFic: page 22. Derek Thomas: 156. D70Focus: pages 99, 173, 180, 190, 204, 218. Dhruvaraj: page 86 below. Ed Seloh: page 142 below. Fisherman's Daughter: pages 3, 130 right, 160, 163, 165, 169, 176, 186, 187, 217. Franco Folini: page 12 left. Frankie Roberto: page 133 above left. Geoff Stearns: page 133 below left. Gilles Gonthier: pages 21, 82 below. Hans Splinter: page 17. Helena40proof: page 59 below. Ian Sutton: pages 2, 152. Indigo Goat: page 15. Jack Wolf: pages 47 below, 73, 76, 80 right, 106, 114. James Emery: page 8 left. J. Baker: page 239. Jeff Turner: page 240. Jen64: page 52 top right. Jessicafm: page: 48. Jessica Merz: page 247. JGD70: page 118. Jilly40uk: pages 5, 78. Joe De Luca: page 195 above. John Bragg: page 27. John Thompson: page 161. Jon Mitchell: pages 37, 59 top left. Judepics: page 44. Justin Beck: page 34. Karen Massier: page 138 right. Kevin Cole: pages 41, 64, 85. Kh.drakkon: page 141 left. Laurent Jégou: page 49. Liz West: page 56. Marvin Smith: page 32. Matthew Folley: page 149. Michael Westhoff: page 150. Mike Baird: page 58 below left. Mindfrieze: page 157. Mr Pbp: page 58 below right. Mullica: page 53. MYK Reeve: page: 134 left. OakleyOriginals: page 23. Orin Zebest: page 237. Paul Sapiano: page 115 Paul Stein: page 93. Phelyan Sanjoin: page 80 left. Plain_Jane53177: page 4. Putneypics: page 125. Neal Jennings: page 18. Rachel Black: page 159. Raphael Guillaumin: page 124. Rich Varney: page 40. Roy Niswanger: page 24. Rumble 1973: pages 198, 211. Smoobs: page 126 left. Spettacolopuro: page 30. Stefan Bungart: page 26. Stefan Kloo: page 11. Steven Wong: page 244. Steve Puntor: page 119. Subberculture: page 54. Syliva McPherson: page 45. Takato Marui: page 57 left. Tambako the Jaguar: pages 25, 101. Tanakawho: page 110. Tastybit: pages 199, 212. The Girls NY: page 19. Thomas Bresson: page 70 right. Tim Gage: page 50 right. Tim Samoff: page 57 below right. Tim Schapker: page 31. Todd Huffman: pages 113, 205, 243. Toholio: page 127. Tomasz Zachariasz: page 38.

ACKNOWLEDGEMENTS

Tony the misfit: page 8 right. Trees for the Future: page 148. Vards UZ vards: page: 134 right. William Warby: pages 61, 100, 112 below. Wobble-san: page: 75. Wohach: page 202. Yuval Haimouits: page 20 right.

Wikimedia Commons:- Aaron Lalz: page 84. Abalg: page: 128 above. Alan Manson: page 142 above left. Alvesgasper: pages 28, 69 below, 71-72, 79 top, 90. Andree Stephan: page 132 above. Bartosz Kosiorek Gang 65: page 89. BKSimonb: page 86 above. Eigene Autnahme: page 195 below. Fir0002: page 81. Jack Dyking: page 74. James D. Ellis: page 233. Jastrow: page 132 below right. Jeff de Longe: pages 68, 68 top right. Joseph G Gall: page 46 left. Gokoenig: page 133 below right. Ken Thomas: page 87. Muhammad Mahdi Karim: page 83. Ohad Balaga: page 77. Pollinator: page 215. Public Domain: pages 65, 70 left, 141 right, 142 right. Robert Engelhardt: pages 173 above, 188 below, 200. Sarefo: page 234. Scott Bauer: page 92. Supermanu: page 39 both. Victor L. Lee: page 9. Waugsberg: pages 88, 94, 95 both, 96, 117, 203, 232 . Wofl: page 69 below right. Wojsyl, 2005: page 145. Wolfgang Sauber: page 132 below left.

iStockphoto LP ©:- Steve Foley (bee logo chapter openers). Alexsol: page 128 below. Annedde: page 164. Arlindo71: page 35. Arne Bramsen: page 139. Bartosz Hadyniak: page 109. Catnap72: pages 155, 172, 196. César Camargo: page 111. Dan Moore: page 226. Darla Hallmark: pages 154, 168. Donald Erickson: page 188-189. Dusan Zidar: pages 121, 221. Eric Delmar: page 185. Gustavo Andrade: page 220. Heinz Waldukat: page 151. Irochka Tischenko: page 97. Ivan Mateev: page 13. Iztok Grilc: page 140. Jakub Niezabitwski: page 122. James Whittaker: pages 16, 162, 178 left. Javier Robles: page 194. Joannawnuk: pages 6-7, 51. Jorg Kaschper: page 144. Jose Juan Garcia: page 170. Joset Philip: page 126 right. Jowita Stachowiak: page 138 left. KKgas: page 166-167. Katrina Brown: page 36. Kelly Cline: page 14. Kenneth Wiedmann: page 183. Liv Friis-Larsen: page 222. Magdalena Kucova: page 223. Marcx: page 131. Martin McCarthy: page 193. Monika Adamczyk: page 224. Kostas Koutsoukos: page 33. Kristian Septimus Krogh: page 179 right. Lloyd Paulson: page 201. Musat: pages: 66-67. Nathaniel Frey: page 60. Noam Armonn: page 197. Olga Langerova: pages 120 right, 228. Pe Jo29: page 227. Peter Engelsted Jonasen: page 81. Phil Berry: pages 50 left, 206. Plainview: page 123. Proximinder: pages 104-105, 112 above, 207, 213, 214. Rachel Giles: page 191. Rebecca Picard: page 225. Robert Milek: page 143. Sergiy Goruppo: page 153. Sergy Babich: page 230-231. Simon Smith: page 174-175. Sondra Paulson: page 120 right. Steven Robertson: page 171. Tedestudio: page 229. Tomasz Szymanski: page 219. Tseon: page 184. William Britten: page 146-147. Will Schmitz: page 42. Xyno: page 137. Yurymelnikau: page 98.